STUDY AND
REVISION GUIDE

Cambridge International
AS & A Level

Physics

Third edition

Richard Woodside
Martin Williams

Boost

HODDER
EDUCATION
AN HACHETTE UK COMPANY

Orders: please contact Hachette UK Distribution, Hely Hutchinson Centre, Milton Road, Didcot, Oxfordshire, OX11 7HH. Telephone: +44 (0)1235 827827. Email education@hachette.co.uk Lines are open from 9 a.m. to 5 p.m., Monday to Friday. You can also order through our website: www.hoddereducation.com

ISBN: 978 1 3983 4440 2

© Richard Woodside, Martin Williams 2022

First published in 2012

Second edition published in 2015

This edition published in 2022 by
Hodder Education,
An Hachette UK Company
Carmelite House
50 Victoria Embankment
London EC4Y 0DZ

www.hoddereducation.com

Impression number   10 9 8 7 6 5 4 3 2

Year   2026 2025 2024

Cover photo © Korn V. - stock.adobe.com

Typeset in India by Integra Software Services Pvt. Ltd, Pondicherry, India

Printed and bound by CPI Grop (UK) Ltd, Croydon, CR0 4YY

A catalogue record for this title is available from the British Library.

Get the most from this book

Everyone has to decide his or her own revision strategy, but it is essential to review your work, learn it and test your understanding. This Study and Revision Guide will help you to do that in a planned way, topic by topic. Use this book as the cornerstone of your revision and don't hesitate to write in it – personalise your notes and check your progress by ticking off each section as you revise.

Tick to track your progress

REVISED ☐

Use the revision planner on page 4 to plan your revision, topic by topic. Tick each box when you have:

» revised and understood a topic
» tested yourself
» practised the exam-style questions and gone online to check your answers

You can also keep track of your revision by ticking off each topic heading in the book. You may find it helpful to add your own notes as you work through each topic.

Features to help you succeed

STUDY TIPS

Tips are given throughout the book to help you polish your exam technique to help maximise your achievement in the exam.

KEY TERMS AND DEFINITIONS

Clear, concise definitions of essential key terms are provided where they first appear and in the glossary.

▶ REVISION ACTIVITIES

Activities to help you understand each topic in an interactive way.

▶ NOW TEST YOURSELF

TESTED ☐

Short, knowledge-based questions provide the first step in testing your learning. Go to **www.hoddereducation.com/cambridgeextras** for the answers.

WORKED EXAMPLES

Working through examples of calculations will help to improve your maths skills and develop your confidence and competence.

PRACTICAL SKILLS

These encourage an investigative approach to the practical work required for your course.

▶ END OF CHAPTER CHECK

Quick-check bullet lists for each topic.

▶ EXAM-STYLE QUESTIONS

Guidance on preparing for examination is followed by example AS questions and A Level questions at the end of each section. Each question has sample answers and comments indicating how the answers could be improved – available at **www.hoddereducation.com/cambridgeextras**.

My revision planner

Countdown to my exams   **5**

Introduction   **6**

REVISED   TESTED   EXAM READY

AS LEVEL

1   Physical quantities and units   8
2   Kinematics   17
3   Dynamics   27
4   Forces, density and pressure   36
5   Work, energy and power   44
6   Deformation of solids   51
7   Waves   56
8   Superposition   66
9   Electricity   76
10   D.C. circuits   83
11   Particle physics   93

AS Level experimental skills and investigations   **101**

AS exam-style questions   **109**

A LEVEL

12   Motion in a circle   115
13   Gravitational fields   120
14   Temperature   127
15   Ideal gases   132
16   Thermodynamics   138
17   Oscillations   140
18   Electric fields   148
19   Capacitance   156
20   Magnetic fields   163
21   Alternating currents   179
22   Quantum physics   184
23   Nuclear physics   194
24   Medical physics   202
25   Astronomy and cosmology   214

A Level experimental skills and investigations   **223**

A Level exam-style questions   **230**

Glossary   **236**

Check your answers at **www.hoddereducation.com/cambridgeextras**

Countdown to my exams

6–8 weeks to go
REVISED ☐

» Start by looking at the syllabus – make sure you know exactly what material you need to revise and the style of the examination. Use the revision planner on page 4 to familiarise yourself with the topics.
» Organise your notes, making sure you have covered everything on the syllabus. The revision planner will help you to group your notes into topics.
» Work out a realistic revision plan that will allow you time for relaxation. Set aside days and times for all the subjects that you need to study, and stick to your timetable.
» Set yourself sensible targets. Break your revision down into focused sessions of around 40 minutes, divided by breaks. This Study and Revision Guide organises the basic facts into short, memorable sections to make revising easier.

2–5 weeks to go
REVISED ☐

» Read through the relevant sections of this book and refer to the study tips, key terms and end of chapter checks. Tick off the topics as you feel confident about them. Highlight those topics you find difficult and look at them again in detail.
» Test your understanding of each topic by working through the 'Now test yourself' questions. Look up the answers at **www.hoddereducation.com/ cambridgeextras**.
» Make a note of any problem areas as you revise, and ask your teacher to go over these in class.
» Look at past papers. They are one of the best ways to revise and practise your exam skills. Check your answers with your teacher. Write or prepare planned answers to the exam-style questions provided in this book and then check your answers against the sample answers and comments at **www.hoddereducation.com/cambridgeextras**.
» Use the revision activities to try different revision methods. For example, you can make notes using mind maps, spider diagrams or flash cards.
» Track your progress using the revision planner and give yourself a reward when you have achieved your target.

1 week to go
REVISED ☐

» Try to fit in at least one more timed practice of an entire past paper and seek feedback from your teacher, comparing your work closely with the mark scheme.
» Check the revision planner to make sure you haven't missed out any topics. Brush up on any areas of difficulty by talking them over with a friend or getting help from your teacher.
» Attend any revision classes put on by your teacher. Remember, teachers are experts at preparing people for examinations.

The day before the examination
REVISED ☐

» Flick through this Study and Revision Guide for useful reminders, for example the study tips, key terms and end of chapter checks.
» Check the time and place of your examination.
» Make sure you have everything you need – extra pens and pencils, tissues, a watch, bottled water, sweets.
» Allow some time to relax and have an early night to ensure you are fresh and alert for the examination.

My exams
REVISED ☐

Paper 1

Date: Time:

Location:

Paper 2

Date: Time:

Location:

Paper 3

Date: Time:

Location:

Paper 4

Date: Time:

Location:

Paper 5

Date: Time:

Location:

Introduction

This revision guide is written to support students following the Cambridge International AS & A Level Physics 9702 course. The assessment of the AS Level is based on examination papers 1 to 3, while at A Level the results from the AS papers are combined with two further papers, Papers 4 and 5. Details of each examination are set out below.

AS Level

Paper 1   1 hour 15 minutes   40 marks

» 40 multiple-choice questions based on the AS Level syllabus content
» 31% of the AS Level, 15.5% of the A Level
» Half the questions test knowledge and understanding; the remainder test handling, applying and evaluating information.

Paper 2   1 hour 15 minutes   60 marks

» Structured questions (usually 7 or 8) based on the AS Level syllabus content
» 46% of AS Level, 23% of A Level

Paper 3   2 hours   40 marks

» Advanced practical skills
» Practical work and structured questions
» There are two questions based on the experimental skills outlined in the practical assessment section of the syllabus.
» 23% of the AS Level, 11.5% of the A Level

A Level

Paper 4   2 hours   100 marks

» Structured questions (usually 12) based on the A Level syllabus content; knowledge of material from AS Level will be required.
» 38.5% of the A Level

Paper 5   1 hour 15 minutes   30 marks

» Planning, analysis and evaluation
» There are two questions based on the experimental skills outlined in the practical assessment section of the syllabus.
» 11.5% of the A Level

Papers 1, 2 and 4 have a data page and a formulae page at the beginning.

Strategies for answering multiple-choice questions

» Read the stem of the question carefully so that you are clear about what the question is asking.
» Read each alternative answer fully, do not just scan for key words.
» Eliminate those answers that you are certain are definitely wrong. This strategy limits the range of answers to select from.
» Link the content of the answer to the key elements in the stem of the question.
» Do not rush. There are 75 minutes to answer 40 questions – it is better to answer 30 questions correctly by using time sensibly rather than 40 questions incorrectly because you have rushed.

Check your answers at **www.hoddereducation.com/cambridgeextras**

» Do not spend too much time on a single question – if you find that you are struggling with a question, move on to the next one and return to the unanswered one later.
» Answer **all** questions, even those where you are unsure what the answer should be – in these cases use your best judgement to select an answer.
» Practise timed questions – use past paper questions to practise your timing.

Answering structured questions

Each question will be subdivided into several parts based around a common theme. At A Level, the subsections may test objectives from different sections of the syllabus. In many structured questions, a value or information found in one part is used in subsequent parts. Half the questions test knowledge and understanding, while the remainder test handling, applying and evaluating information. For example, the first part of a question may require you to state a definition or write an equation. This will be followed by a description, often with some data. At AS Level, you will be expected to use the data along with the definition or equation from the first part of the question. At A Level, you might be required to make short written statements including descriptions or explanations, carry out calculations, sketch a graph or draw a simple diagram.

AS Level practical assessment

There are two questions of 1 hour each. For each question, there will be some apparatus available and instructions about how the apparatus is to be put together, followed by instructions about what readings to take.

The mark distribution for each skill tested varies according to the question:

Skill	Question 1	Question 2
Manipulation, measurement and observation	7	5
Presentation of data and observations	6	2
Analysis, conclusions and evaluation	4	10

In each case, the examiner will allocate the remaining 3 marks across the skills. The paper will not test your knowledge and understanding of any of the subject content. See pages 101–108 for further details.

A Level practical assessment

There are two questions in paper 5, each worth 15 marks (see also pages 223–229). The first question requires you to plan an experiment to investigate an aspect of physics given in the question. In the second question, you will be given data to analyse and evaluate. Each question is designed to test different practical skills:

Skill	Question 1	Question 2
Planning	15	–
Analysis, conclusions and evaluation	–	15

Command words

Command words are used in examination questions to help you understand what is expected in the answer. Check your syllabus and make sure you know what each command word requires you to do. The syllabus is available on the Cambridge International website at **www.cambridgeinternational.org**.

1 Physical quantities and units

SI units

Base quantities

All quantities in science consist of a numerical magnitude and a unit. **SI units** are based on the units of seven SI **base quantities**, of which you need to be familiar with five:

»» mass – kilogram (kg)
»» length – metre (m)
»» time – second (s)
»» temperature – kelvin (K)
»» electric current – ampere (A)

Although it is not formally an SI unit, the degree Celsius (°C) is often used as a measure of temperature.

Each of these units has a precise definition. You do not need to remember the details of these definitions.

KEY TERMS

SI units (Système International d'Unités) are carefully defined units that are used throughout the scientific world for measuring all quantities.

Base quantities are fundamental quantities whose units are used to derive all other units.

Derived units

»» The units of all other quantities are derived from the **base units**.
»» For example, speed is found by dividing the distance travelled by the time taken. Therefore, the unit of speed is metres (m) divided by seconds (s).
»» At O Level or IGCSE, you will probably have written this unit as m/s. Now that you are taking your studies a stage further, you should write it as $m\,s^{-1}$.

KEY TERMS

Base units are the units of the base quantities.

Derived units are combinations of base units.

WORKED EXAMPLE

The unit of force is the newton. What is this in base SI units?

Answer

The newton is defined from the equation:

force = mass × acceleration

unit of mass = kg

unit of acceleration = $m\,s^{-2}$

Insert into the defining equation:

units of newton = kg × m × s^{-2} or $kg\,m\,s^{-2}$

▶ NOW TEST YOURSELF

1 The quantity power is a derived quantity and its unit, the watt (W), is a derived unit. Express watts in base units.
2 Which of the following are base quantities?
   time, speed, volume, energy
3 Which of the following are base units?
   kilogram, metre squared, joule, kelvin

STUDY TIP

To break watts down into base units, use the format $J\,s^{-1}$; then substitute in for J and continue using positive and negative indices.

Homogeneity of equations

REVISED

If you are not sure whether an equation is correct, you can use the units of the different quantities to check it. The units on both sides of the equation must be the same.

WORKED EXAMPLE

When an object falls in a vacuum, all its gravitational potential energy is converted into kinetic energy. By comparing units, show that the equation $mg\Delta h = \frac{1}{2}mv^2$ is a possible solution to this problem.

Answer

Write down the units of the quantities on each side of the equation.

Left-hand side: unit of m = kg; unit of g = ms^{-2}; unit of h = m

Right-hand side: unit of $\frac{1}{2}$ = none; unit of m = kg; unit of v = ms^{-1}

Compare the two sides:

units of $mg\Delta h$ = kg $\times$ ms^{-2} $\times$ m = $kg\,m^2s^{-2}$

units of $\frac{1}{2}mv^2$ = kg $\times$ $(ms^{-1})^2$ = $kg\,m^2s^{-2}$

Both sides of the equation are identical.

> **STUDY TIP**
>
> Try to work through the worked examples before reading the answer and then compare your answer with the one supplied.

▶ NOW TEST YOURSELF

TESTED

4 The pressure exerted beneath the surface of a liquid is given by the equation:

$$\Delta p = \rho g \Delta h$$

where p is pressure, h is depth below the surface, ρ is density of the liquid and g is the acceleration due to gravity. Show that the equation is homogeneous.

5 When a sphere falls through a liquid, there is a drag force on it. Stokes' law states that the drag force F is given by the formula:

$$F = 6\pi\eta rv$$

where F is force (N), η is viscosity of the fluid $(kg\,m^{-1}s^{-1})$, r is radius of the sphere (m) and v is relative velocity between the fluid and sphere (ms^{-1}). Show that the units in this formula are homogeneous.

Using standard form

REVISED

One way to deal with very large or very small quantities is to use standard form. Here, the numerical part of a quantity is written as a single digit followed by a decimal point, and as many digits after the decimal point as are justified; this is then multiplied by 10 to the required power.

WORKED EXAMPLE

a The output from a power station is $5\,600\,000\,000\,W$. Express this in watts, using standard form.

b The charge on an electron is $0.000\,000\,000\,000\,000\,000\,16\,C$. Express this in standard form.

Answer

a $5\,600\,000\,000\,W = 5.6 \times 10^9\,W$

b $0.000\,000\,000\,000\,000\,000\,16\,C = 1.6 \times 10^{-19}\,C$

'$\times\,10^{-19}$' means that the number, in this case 1.6, is divided by 10^{19}.

> **STUDY TIP**
>
> Mistakes are often made when dividing by numbers in standard form. If you are dividing by a quantity like 1.6×10^{-19}, remember that $\frac{1}{10^{-x}} = 10^{+x}$.

An added advantage of using standard form is that it indicates the degree of precision to which a quantity is measured. This will be looked at in more detail in 'AS Level experimental skills and investigations' on pp. 101–108.

Multiples and submultiples of base units

REVISED

Sometimes, the base unit is either too large or too small. Prefixes are used to alter the size of the unit. Table 1.1 shows the prefixes that you need to know.

▼ Table 1.1

Prefix	Symbol	Meaning	
pico	p	$\div 1\,000\,000\,000\,000$	$\times 10^{-12}$
nano	n	$\div 1\,000\,000\,000$	$\times 10^{-9}$
micro	µ	$\div 1\,000\,000$	$\times 10^{-6}$
milli	m	$\div 1000$	$\times 10^{-3}$
centi	c	$\div 100$	$\times 10^{-2}$
deci	d	$\div 10$	$\times 10^{-1}$
kilo	k	$\times 1000$	$\times 10^{3}$
mega	M	$\times 1\,000\,000$	$\times 10^{6}$
giga	G	$\times 1\,000\,000\,000$	$\times 10^{9}$
tera	T	$\times 1\,000\,000\,000\,000$	$\times 10^{12}$

These are the recognised SI prefixes. The deci- (d) prefix is often used in measuring volume – decimetre cubed (dm^3) is particularly useful.

STUDY TIP

Remember that $1\,dm^3$ is $\frac{1}{1000}$ (not $\frac{1}{10}$) of $1\,m^3$. It is really $(dm)^3$.

Hence, it is $\frac{1}{10}m \times \frac{1}{10}m \times \frac{1}{10}m$.

▶ **NOW TEST YOURSELF**　　　　　　TESTED

6 The distance from the Earth to the Sun is $150\,Gm$. Express this distance in metres using standard form.

7 The wavelength of violet light is about $4 \times 10^{-7}\,m$. Express this wavelength in nanometres.

8 Calculate the number of micrograms in a kilogram. Give your answer in standard form.

Making estimates of physical quantities

REVISED

There are a number of physical quantities where you should be aware of the rough values, for example, the speed of sound in air ($\approx 300\,m\,s^{-1}$). Lists of such values are given in appropriate parts of this study and revision guide – an example is Table 7.2 on p. 62.

Errors and uncertainties

Errors

REVISED

An **error** is a mistake in a reading caused by either faulty apparatus or poor technique. Errors and repeated readings are discussed in more detail on p. 106.

Accuracy, precision and uncertainty

Accuracy and **precision** are terms that often cause confusion.

Consider a rod of 'true' diameter 52.8012 mm. Suppose that you use a ruler and measure it to be 53 mm. This is accurate, but it is not very precise. If your friend uses a micrometer screw gauge and measures it as 52.81 mm, this is more precise, even though the final figure is not totally accurate.

No measurement can be made to absolute precision – there is always some **uncertainty**.

If a result is recorded as 84.5 s, this implies that there is an uncertainty of at least 0.1 s, perhaps more. You might see such a reading written as 84.5 ± 0.2 s. The 0.2 s in this reading is called the **absolute uncertainty**.

It is often convenient to express an uncertainty as a percentage of the reading. This is known as the **percentage uncertainty**.

$$\text{percentage uncertainty} = \frac{\text{absolute uncertainty}}{\text{reading}} \times 100\%$$

> **KEY TERMS**
>
> **Accuracy** is how close to the 'real value' a measurement is.
>
> **Precision** is the part of accuracy that the experimenter controls by the choice of measuring instrument and the skill with which it is used. It refers to how close a set of measured values are to each other.
>
> **Uncertainty** is the range of values in which a measurement can fall.

WORKED EXAMPLE

A technician records the length of a rod as 84.5 ± 0.2 m. Determine the percentage uncertainty in this measurement.

Answer

$$\frac{0.2}{84.5} \times 100\% = 0.24\%$$

Precision of measurement

» When making a static measurement (for example, the length of a pendulum) you should normally measure to the nearest division on the instrument.
» If the divisions are 1 millimetre or more apart, you need to judge to the nearest half division or better.
» When making a dynamic measurement (for example, the height to which a ball bounces), then other considerations come into play – the ball is moving, so you have to judge when it is at its maximum height. This is a much more difficult task. You can probably only measure this to the nearest 5 millimetres.
» Many digital stopwatches measure to 1/100 of a second. However, the uncertainties in the reaction times of manually starting and stopping a stopwatch are much greater than this. The best you can manage is to measure to the nearest 1/10 of a second.
» Until 1977, world records for running events were given to a precision of 0.1 s.
» It was only with the advent of electronic timing that it became possible to record race times to 1/100 of a second.
» The current world record for the men's 100 m is 9.58 s. This suggests an absolute uncertainty of ±0.01 s, and a percentage uncertainty of approximately 0.1%.
» The knock-on effect is that, for a world record to be valid, the track length must also be measured to a precision of 0.1% or better, which is an absolute uncertainty of 10 cm.

Repeat readings

» Precision can also be estimated from taking repeat readings.
» For example, when five readings are taken of the time for a ball to run down a track, it is acceptable to give the uncertainty as half the range of the readings.

WORKED EXAMPLE

The time of flight for a projectile is measured by five technicians. The readings are:

$5.2\,s, 5.2\,s, 5.4\,s, 5.0\,s, 5.1\,s$

Calculate the absolute uncertainty in these readings.

Answer

The range is the difference between the largest and smallest values ($5.4 - 5.0 = 0.4\,s$).

The absolute uncertainty is $\pm \dfrac{0.4\,s}{2} = \pm 0.2\,s$.

► NOW TEST YOURSELF

9 What is the realistic absolute uncertainty of a time measured using a hand-held stopwatch?
10 A cylinder is machined to a diameter of 40.24 mm. Four apprentices are asked to measure the diameter of the cylinder using different instruments. The results are shown in Table 1.2.
   Which apprentice gives:
   a the most precise result
   b the most accurate result?

▼ Table 1.2

Apprentice	Diameter
A	4 cm
B	40 mm
C	40.01 mm
D	40.2 mm

Combining uncertainties

To find the uncertainty of a combination of variables, the rules are as follows:

» For quantities that are added or subtracted, the *absolute* uncertainties are added.
» For quantities that are multiplied together or divided, the *percentage* (or *fractional*) uncertainties are added.
» For a quantity that is raised to a power, to calculate a final uncertainty, the percentage uncertainty is multiplied by the power and the result is treated as a *positive* uncertainty.

WORKED EXAMPLE

The currents coming into a junction are I_1 and I_2. The current coming out of the junction is I. In an experiment, the values of I_1 and I_2 are measured as $2.0 \pm 0.1\,A$ and $1.5 \pm 0.2\,A$, respectively.

Write down the value of I with its uncertainty.

Answer

$$I = I_1 + I_2 = (2.0 \pm 0.1) + (1.5 \pm 0.2)$$

The quantities are being added so to find the uncertainty, the uncertainties of the original quantities are added.

Hence:

$$I = 3.5 \pm 0.3\,A$$

Check your answers at **www.hoddereducation.com/cambridgeextras**

WORKED EXAMPLE

The acceleration of free fall g is determined by measuring the period of oscillation T of a simple pendulum of length L. The relationship between g, T and L is given by the formula:

$$g = 4\pi^2\left(\frac{L}{T^2}\right)$$

In the experiment, L was measured as $0.55 \pm 0.02\,\text{m}$ and T as $1.50 \pm 0.02\,\text{s}$.

Find the value of g and its uncertainty.

Answer

$$g = 4\pi^2\left(\frac{L}{T^2}\right) = 4\pi^2\left(\frac{0.55}{1.50^2}\right) = 9.7\,\text{m s}^{-2}$$

To find the uncertainties, the second and third rules are applied.

percentage uncertainty in $L = \left|\frac{0.02}{0.55}\right| \times 100 = 3.6\%$

percentage uncertainty in $T = \left|\frac{0.02}{1.50}\right| \times 100 = 1.3\%$

percentage uncertainty in $T^{-2} = 2 \times 1.3 = 2.6\%$

percentage uncertainty in g = percentage uncertainty in L + percentage uncertainty in $T^{-2} = 3.6 + 2.6 = 6.2\%$

absolute uncertainty in $g = 9.7 \times \frac{6.2}{100} = 0.6$

Thus:

$$g = 9.7 \pm 0.6\,\text{m s}^{-2}$$

It is worth noting that in this example the question is asking for the absolute uncertainty, not the percentage uncertainty. If you take the short cut and leave your answer as $9.7 \pm 6.2\%$, you will lose credit.

> **PRACTICAL SKILL**
>
> It is also worth noting that it is poor experimental practice to take only one reading and to try to find a value of g from that. You should take a series of readings of T for different lengths L, and then plot a graph of T^2 against L. The gradient of this graph would be equal to $4\pi^2/g$. For further information about graphs see 'Rules for plotting graphs' on pp. 103–104.

Scalars and vectors

Scalar quantities and vector quantities

REVISED ☐

>> A **scalar quantity** has magnitude only. Examples are mass, volume and energy.
>> A **vector quantity** has magnitude and direction. Examples are force, velocity and acceleration.

Adding scalars

>> Consider two masses of 2.4 kg and 5.2 kg.
>> The total mass is 7.6 kg. This total is simply the arithmetic total.

Adding vectors

>> When vectors are added, their directions must be taken into account.
>> Two forces of 3 N and 5 N acting in the same direction would give a total force of 8 N.
>> If two forces of 3 N and 5 N act in opposite directions, the total force is $(5 - 3)\,\text{N} = 2\,\text{N}$, in the direction of the 5 N force.
>> If the two forces act at any other angle to each other, a vector diagram, known as the **vector triangle** is used (see p. 14).
>> If there are more than two vectors, these can be added in a similar manner using a polygon rather than a triangle, provided that all the vectors are **coplanar**, which means that they all lie in the same plane.

> **KEY TERMS**
>
> A set of vectors are said to be **coplanar** if all the vectors in the set lie in the same plane.

Constructing a vector diagram

- » In a vector diagram, each vector is represented by a line.
- » The magnitude of the vector is represented by the length of the line and its direction is represented by the direction of the line.
- » If two vectors act at a point, their resultant can be found by drawing a vector triangle.

The following rules will help you to draw a vector triangle (Figure 1.1):

1 Choose a suitable scale. Draw a line to represent the first vector (V_1) in both magnitude and direction.
2 Draw a second line, starting from the tip of the first line, to represent the second vector (V_2) in both magnitude and direction.
3 Draw a line from the beginning of the first vector to the end of the second vector to complete a triangle.
4 The resultant vector is represented by the length of this line and its direction.

> **STUDY TIP**
>
> The larger the scale you choose, the greater precision you should achieve in your answer. It is good practice to include your scale on the diagram. When measuring distances, use a ruler and when measuring angles, use a protractor.

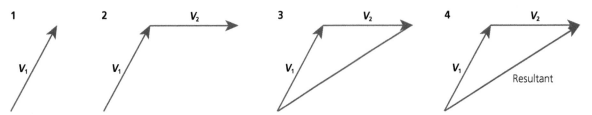

▲ **Figure 1.1 Drawing a vector triangle**

WORKED EXAMPLE

A light aircraft is flying due north with a velocity relative to the air of $200\,\text{km}\,\text{h}^{-1}$. A wind from 30° north of west starts to blow at $80\,\text{km}\,\text{h}^{-1}$ (Figure 1.2). Calculate the velocity of the aircraft relative to the ground.

▲ **Figure 1.2**

Answer

Draw a vector diagram to a scale of $1.0\,\text{cm}:40\,\text{km}\,\text{h}^{-1}$ (Figure 1.3).

> length of the resultant $= 4.35\,\text{cm}$

Multiply by the scaling:

> velocity $= 4.35 \times 40\,\text{km}\,\text{h}^{-1} = 174\,\text{km}\,\text{h}^{-1}$

Measure the angle θ, using a protractor:

> $\theta = 23°$, so the direction is 23° east of north

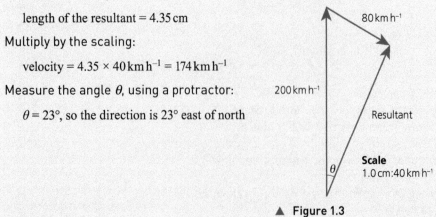

▲ **Figure 1.3**

> **STUDY TIP**
>
> Compare the layout of your answer with Figure 1.3. Is your layout clear? Can the examiner see what you have tried to do? This is important in calculations, where some credit will be given even if an arithmetic error leads to you giving the wrong answer.

You will observe that the directions of the original two vectors go round the triangle in the same direction (in this example clockwise). The direction of the resultant goes in the opposite direction (anticlockwise).

Check your answers at **www.hoddereducation.com/cambridgeextras**

Components at right angles: resolving vectors

Just as it is useful to be able to combine vectors, it is also useful to be able to resolve vectors into components at right angles to each other.

Figure 1.4 shows a vector, **V**, acting at an angle θ to the horizontal.

The vector triangle in Figure 1.4(a) shows that this vector can be considered to be made up from a vertical component (V_v) and a horizontal component (V_h). It is sometimes easier to use a diagram similar to Figure 1.4(b) when resolving vectors – this emphasises that the vectors are acting at the same point.

By inspection you can see that $\cos\theta = V_h/V$. Therefore:

$$V_h = V\cos\theta$$

Similarly:

$$V_v = V\sin\theta$$

(a)
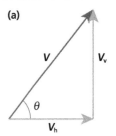

(b)
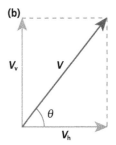

▲ Figure 1.4

WORKED EXAMPLE

A box of weight 20 N lies at rest on a slope, which is at 30° to the horizontal. Calculate the frictional force on the box up the slope.

Answer

Resolve the weight (20 N) into components parallel to and perpendicular to the slope (Figure 1.5).

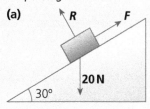

▲ Figure 1.5

The frictional force, F, is equal to the component of the weight down the slope:

$$F = 20\sin30 = 10\,\text{N}$$

11 Two forces, one of magnitude 1200 N and the other of 1000 N, are applied to an object. The angle between the forces is 40°.
Draw a vector diagram and determine the magnitude of the resultant force and the angle it makes with the 1200 N force.

12 A rope is pulled so that it makes an angle of 20° with the vertical.
The tension in the rope is 500 N. Resolve the tension force into vertical and horizontal components.

► REVISION ACTIVITY

Work with a partner. One person gives a key term from this chapter and the other gives an explanation of the term. Change roles until all the key terms have been covered. Do this at the end of every chapter.

END OF CHAPTER CHECK

In this chapter, you have learnt:
- ▶ that all physical quantities consist of a numerical magnitude and a unit ☐
- ▶ to recall the SI base quantities and their units: mass (kg), length (m), time (s), current (A) and temperature (K) ☐
- ▶ to express derived units as products or quotients of the base units ☐
- ▶ to use the named units listed in the syllabus as appropriate ☐
- ▶ to use SI units to check the homogeneity of physical equations ☐
- ▶ to understand and use standard form ☐
- ▶ to recall and use the prefixes and their symbols to indicate decimal submultiples and multiples of base and derived units: pico (p), nano (n), micro (μ), milli (m), centi (c), deci (d), kilo (k), mega (M), giga (G), tera (T) ☐
- ▶ to make reasonable estimates of physical quantities included in the syllabus ☐

- ▶ to understand and explain the effects of systematic errors (including zero errors) and random errors in measurements ☐
- ▶ to understand the distinction between precision and accuracy ☐
- ▶ to assess the uncertainty in a derived quantity by simple addition of absolute or percentage uncertainties ☐
- ▶ to understand the difference between vector and scalar quantities ☐
- ▶ to give examples of scalars and vectors ☐
- ▶ to construct vector diagrams and determine unknown vectors from the diagrams ☐
- ▶ to understand that coplanar vectors are sets of vectors in which all the vectors lie in the same plane ☐
- ▶ to add and subtract coplanar vectors ☐
- ▶ to represent a vector as two perpendicular components ☐

Check your answers at **www.hoddereducation.com/cambridgeextras**

2 Kinematics

Equations of motion

Definitions of quantities

You should know the definitions of the terms **distance**, **displacement**, **speed**, **velocity** and **acceleration**.

> **KEY TERMS**
>
> **Distance** is the length between two points measured along a straight line joining the two points.
>
> **Displacement** is the distance of an object from a fixed reference point in a specified direction.
>
> **Speed** is the distance travelled per unit time.
>
> **Velocity** is the change in displacement per unit time.
>
> **Acceleration** is the rate of change of velocity.

» Distance is a scalar quantity. It has magnitude only.
» Displacement is a vector quantity. It has both magnitude and direction.
» Speed is a scalar quantity. It refers to the total distance travelled.
» Velocity is a vector quantity, being derived from displacement – not the total distance travelled.
» Acceleration is a vector quantity. Acceleration in the direction in which an object is travelling will increase its velocity. Acceleration in the opposite direction from which an object is travelling will decrease its velocity. Acceleration at an angle of 90° to the direction an object is travelling in will change the direction of the velocity but will not change the magnitude of the velocity.

Equations linking the quantities

$$v = \frac{\Delta s}{\Delta t}$$

where v is the velocity and Δs is the change of displacement in time Δt.

$$a = \frac{\Delta v}{\Delta t}$$

where a is the acceleration and Δv is the change in velocity in time Δt.

> **STUDY TIP**
>
> In general, the symbol Δ means 'change of', so Δs is the change in displacement and Δt is the change in time.

Units

Speed and velocity are measured in metres per second ($m\,s^{-1}$).

Acceleration is the change in velocity per unit time. Velocity is measured in metres per second ($m\,s^{-1}$) and time is measured in seconds (s), which means that acceleration is measured in metres per second every second ($m\,s^{-1}$ per s) which is written as $m\,s^{-2}$.

The following worked example demonstrates the difference between speed and velocity.

WORKED EXAMPLE

A toy train travels round one circuit of a circular track of circumference 2.4 m in 4.8 s. Calculate:

a the average speed

b the average velocity

Answer

a If x is the distance travelled:

$$\text{average speed} = \frac{\Delta x}{\Delta t} = \frac{2.4\ (m)}{4.8\ (s)} = 0.50\,\text{m s}^{-1}$$

b s is the displacement, which after one lap is zero. The train finishes at the same point at which it started. Hence:

$$\text{average velocity, } v = \frac{\Delta s}{\Delta t} = \frac{0\ (m)}{4.8\ (s)} = 0\,\text{m s}^{-1}$$

STUDY TIP

It is good practice to include units in your calculations, as shown in this example – it can help you to avoid mistakes with multiples of units. It can also help you to see if an equation does not balance. In most cases in this guide, in order to make the equation clear, units are only included in the final quantity.

Here is a second worked example which demonstrates the relationships between distance, velocity and acceleration.

WORKED EXAMPLE

A car travels 840 m along a straight, level track at a constant speed of 35 m s^{-1}. The driver then applies the brakes and the car decelerates to rest at a constant rate in a further 7.0 s. Calculate:

a the time for which the car is travelling at a constant velocity

b the acceleration of the car when the brakes are applied

Answer

a $v = \dfrac{\Delta s}{\Delta t}$ $35 = \dfrac{840}{\Delta t}$ $\Delta t = \dfrac{840}{35}$

$\Delta t = 24\,\text{s}$

b $a = \dfrac{\Delta v}{\Delta t} = \dfrac{0 - 35}{7.0}$

$a = -5.0\,\text{m s}^{-2}$

STUDY TIP

The minus sign shows that the velocity decreases rather than increases. It is also worth noting that the given quantities in the question are to two significant figures. Therefore, the answer should also be recorded to two significant figures.

> ### NOW TEST YOURSELF
>
> TESTED ☐
>
> 1 Which of the following are vector quantities?
> distance, displacement, speed, velocity, acceleration
> 2 A cyclist starting from rest accelerates at a constant 1.2 m s^{-2}.
> Calculate how long it will take her to reach a speed of 7.8 m s^{-1}.
> 3 A car travelling at 15 m s^{-1} applies its brakes and comes to rest after 4.0 s.
> Calculate the acceleration of the car.

Graphs

Graphs give a visual representation of the manner in which one variable changes with another. Looking at motion graphs can help us to see what is happening over a period of time.

Displacement–time graphs

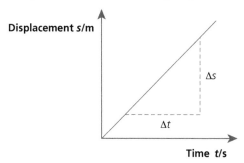

▲ Figure 2.1 Displacement–time graph for constant velocity

Figure 2.1 shows the displacement of an object that increases uniformly with time. This shows constant velocity. The magnitude of the velocity is equal to the gradient of the graph.

$$v = \text{gradient} = \frac{\Delta s}{\Delta t}$$

> **STUDY TIP**
>
> When you measure the gradient of a straight-line graph, use as much of the straight line as possible. This will reduce the percentage error in your calculation.

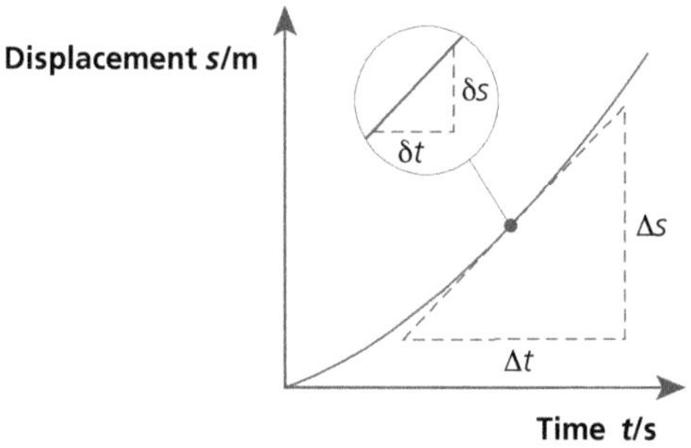

▲ Figure 2.2 Displacement–time graph for increasing velocity

Figure 2.2 shows an example of an object's velocity steadily increasing with time. To find the velocity at a particular instant (the instantaneous velocity) we draw a tangent to the graph at the relevant point and calculate the gradient of that tangent.

Velocity–time graphs

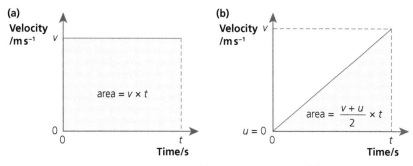

▲ Figure 2.3 Velocity–time graphs: (a) constant velocity, (b) velocity increasing at constant rate

» Figure 2.3(a) shows an object moving with a constant velocity.
» Figure 2.3(b) shows that the velocity of the object is increasing from rest at a constant rate – it has constant acceleration.

The gradient of a velocity–time graph is the change in velocity divided by the time taken. It is equal to the magnitude of the acceleration.

$$a = \frac{v - u}{t_2 - t_1} = \frac{\Delta v}{\Delta t}$$

Displacement from a velocity–time graph

The displacement is equal to the area under a velocity–time graph. This can be clearly seen in Figure 2.3(a). The shaded area is a rectangle and its area is equal to:

height × length = velocity × time

Figure 2.3(b) shows changing velocity; the distance travelled is the average velocity multiplied by the time. For constant acceleration from zero velocity, this is half the maximum velocity multiplied by the time – the area of the shaded triangle.

WORKED EXAMPLE

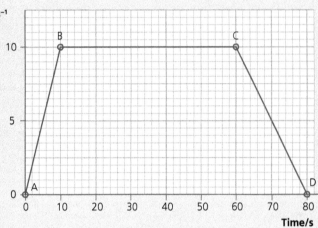

▲ **Figure 2.4**

Figure 2.4 shows the motion of a cyclist as he travels from one stage to the next in a race. Calculate:

a the acceleration from A to B
b the maximum velocity of the cyclist
c the total distance the cyclist travels
d the acceleration from C to D

Answer

a acceleration = gradient = $\dfrac{10 - 0}{10 - 0}$ = 1.0 m s^{-2}

b The maximum velocity can be read directly from the graph. It is 10 m s^{-1}.

c distance travelled = area under the graph
= (½ × 10 × 10) + (10 × 50) + (½ × 10 × 20)
= 650 m

d acceleration = gradient = $\dfrac{0 - 10}{80 - 60}$ = −0.50 m s^{-2}

NOW TEST YOURSELF

TESTED ☐

4 In a test run, a motorcyclist starts from rest and accelerates at 4.4 m s^{-2} for 7.5 s. She then continues at a constant velocity for 20 s before applying the brakes and coming to rest in a further 3.0 s.

a Calculate the motorcyclist's maximum velocity.
b Draw a velocity–time graph of the test run.
c From the graph, determine the deceleration of the motorcyclist in the last part of the journey. (Assume that the deceleration is constant.)
d Use the graph to determine the total distance the motorcyclist travels.

Check your answers at **www.hoddereducation.com/cambridgeextras**

Deriving equations of uniformly accelerated motion

Figure 2.5 shows the motion of an object that has accelerated at a uniform rate, from an initial velocity u to a final velocity v in time t.

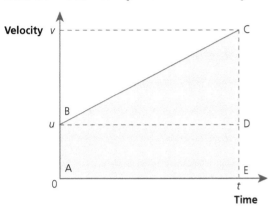

▲ Figure 2.5

Equation 1

The acceleration of the object:

$$a = \frac{\Delta v}{\Delta t} = \frac{v - u}{t}$$

Rearranging this equation gives:

$$v = u + at$$

Equation 2

The distance s travelled by the object can be found in two ways:

$$s = \text{average velocity} \times \text{time} = \frac{v + u}{2}t$$

Equation 3

The distance s travelled is equivalent to the area under the graph:

$$s = \text{area of rectangle ABDE} + \text{area of triangle BCD}$$

$$s = ut + \tfrac{1}{2}(v - u)t$$

From Equation 1:

$$\frac{v - u}{t} = a$$

Therefore:

$$s = ut + \tfrac{1}{2}at^2$$

Equation 4

A fourth equation is needed to solve problems in which the time and one other variable are not known.

Equation 1 rearranges to:

$$t = \frac{v - u}{a}$$

Substitute this in Equation 2:

$$s = \frac{v + u}{2} \times \frac{v - u}{a} = \frac{v^2 - u^2}{2a}$$

Rearranging gives:

$$v^2 = u^2 + 2as$$

STUDY TIP

These equations of motion can only be used if there is constant acceleration (including constant deceleration and zero acceleration) for the whole part of the journey that is being considered.

Summary

The equations of uniformly accelerated motion are:

$$v = u + at \qquad\qquad s = ut + \tfrac{1}{2}at^2$$

$$s = \frac{v + u}{2}t \qquad\qquad v^2 = u^2 + 2as$$

Using the equations of uniformly accelerated motion

A common type of problem you might be asked to analyse is the journey of a vehicle between two fixed points.

WORKED EXAMPLE

During the testing of a car, it is timed over a measured kilometre. In one test, it enters the timing zone at a velocity of $50\,\mathrm{m\,s^{-1}}$ and decelerates at a constant rate of $0.80\,\mathrm{m\,s^{-2}}$. Calculate:

a   the velocity of the car as it leaves the measured kilometre

b   the time it takes to cover the measured kilometre

Answer

a $u = 50\,\mathrm{m\,s^{-1}}$

   $s = 1.0\,\mathrm{km} = 1000\,\mathrm{m}$

   $a = -0.80\,\mathrm{m\,s^{-2}}$

   $v = ?$

   Required equation:

   $$v^2 = u^2 + 2as$$

   Substitute the relevant values and solve the equation:

   $$v^2 = 50^2 + 2 \times (-0.80) \times 1000 = 2500 - 1600 = 900$$

   $$v = 30\,\mathrm{m\,s^{-1}}$$

b   Required equation:

   $$v = u + at$$

   Substitute in the relevant variables:

   $$30 = 50 - (0.80 \times t)$$

   $$t = \frac{(50 - 30)}{0.8}$$

   $$= 25\,\mathrm{s}$$

> **STUDY TIP**
>
> It might seem tedious to write out all the quantities you know and the equation you are going to use. However, this will mean that you are less likely to make a careless error and, if you do make an arithmetic error, it helps the examiner to see where you have gone wrong, so that some marks can be awarded.

> **STUDY TIP**
>
> Two common mistakes in this type of question are:
> » forgetting that deceleration is a negative acceleration
> » forgetting to convert kilometres to metres

▶ NOW TEST YOURSELF

5   An athlete starts from rest and accelerates uniformly for $3.0\,\mathrm{s}$ at a rate of $4.2\,\mathrm{m\,s^{-2}}$. Calculate the athlete's final speed.

6   A car on a test run enters a measured kilometre at a velocity of $28\,\mathrm{m\,s^{-1}}$. It leaves the measured kilometre at a velocity of $40\,\mathrm{m\,s^{-1}}$. Calculate:

   a   the time between the car entering and leaving the measured kilometre

   b   the average acceleration of the car

Analysing the motion of an object in a uniform gravitational field

REVISED

The equations of uniformly accelerated motion can be used to analyse the motion of an object moving vertically under the influence of gravity. In this type of example, it is important to call one direction positive and the other negative and to be consistent throughout your calculation. The following worked example demonstrates this.

WORKED EXAMPLE

A boy throws a stone vertically up into the air with a velocity of $6.0\,\mathrm{m\,s^{-1}}$. The stone reaches a maximum height and falls into the sea, which is 12 m below the point of release (Figure 2.6). Calculate the velocity at which the stone hits the water surface. (acceleration due to gravity = $9.81\,\mathrm{m\,s^{-2}}$)

Answer

$u = 6.0\,\mathrm{m\,s^{-1}}$

$a = -9.81\,\mathrm{m\,s^{-2}}$

$s = -12\,\mathrm{m}$

$v = ?$

Required equation:

$v^2 = u^2 + 2as$

$v^2 = 6.0^2 + [2 \times (-9.81) \times (-12)] = 36 + 235 = 271$

$v = \pm 16.47\,\mathrm{m\,s^{-1}}$

In this example, upwards has been chosen as the positive direction; hence, u is $+6.0\,\mathrm{m\,s^{-1}}$. Consequently, the distance of the sea below the point of release (12 m) and the acceleration due to gravity ($9.81\,\mathrm{m\,s^{-2}}$) are considered negative because they are both in the downward direction.

The final velocity of the stone is also in the downward direction. Therefore, it should be recorded as $-16.47\,\mathrm{m\,s^{-1}}$ and rounded to $-16\,\mathrm{m\,s^{-1}}$.

It is also worth noting that air resistance on a stone moving at these speeds is negligible and can be ignored.

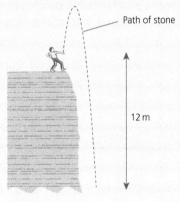

Path of stone

12 m

▲ **Figure 2.6**

PRACTICAL SKILL

When timing moving objects, readings must always be repeated and then averaged. One method of finding the uncertainty in timing is to halve the difference between the maximum and minimum readings. For further information on determining uncertainties, see 'Evaluation' on pp. 105–107.

Acceleration of free fall

REVISED

In the absence of air resistance, all objects near the Earth fall with the same acceleration. This is known as the acceleration of free fall. Similarly, objects near any other planet will fall with equal accelerations. However, these accelerations will be different from those near the Earth. This is explored further in the section on dynamics.

Determination of the acceleration of free fall

Figure 2.7 shows apparatus that can be used to determine the acceleration of free fall.

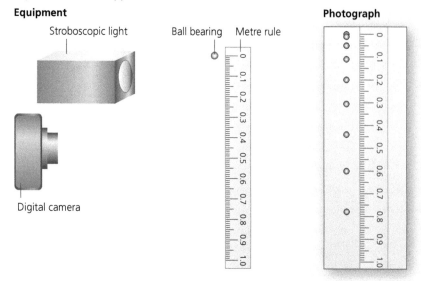

▲ **Figure 2.7 Apparatus to determine the acceleration of free fall**

The stroboscopic light flashes at a fixed frequency and the shutter of the camera is held open. This results in a photograph that shows the position of the ball in successive time intervals, as in Figure 2.7. In this example, the stroboscopic light was set to flash at 20 Hz. In Table 2.1, the third column shows the distance the ball travels in each time interval and the fourth column shows the average speed during each interval.

▼ **Table 2.1**

Time/s	Position/m	Distance/m	Speed/m s^{-1}
0.00	0.00	0.00	0.0
0.05	0.01	0.01	0.2
0.10	0.05	0.04	0.8
0.15	0.11	0.06	1.2
0.20	0.20	0.09	1.8
0.25	0.30	0.11	2.2
0.30	0.44	0.14	2.8
0.35	0.60	0.16	3.2
0.40	0.78	0.18	3.6

A graph of the displacement against time is plotted (Figure 2.8). Acceleration is equal to the gradient of this graph.

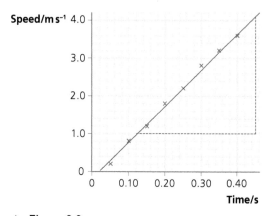

▲ **Figure 2.8**

Check your answers at **www.hoddereducation.com/cambridgeextras**

Readings from Figure 2.8: (0.45, 4.1) and (0.125, 1.0).

$$a = \frac{4.1 - 1.0}{0.45 - 0.125} = 9.5\,\text{m}\,\text{s}^{-2}$$

> ## NOW TEST YOURSELF
> TESTED ☐
>
> 7 An astronaut standing on the Moon drops a hammer from a height of 1.2 m. The hammer strikes the ground 1.2 s after being released. Calculate the acceleration of free fall on the Moon.

Motion in two dimensions
REVISED ☐

Consider an object thrown from near the Earth's surface with an initial velocity v at an angle θ to the horizontal. The velocity v can be resolved into two components, one horizontal and one vertical, as can be seen in Figure 2.9.

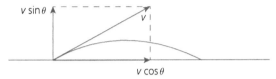

▲ **Figure 2.9 Motion of an object in a gravitational field can be resolved into components in the vertical and horizontal directions**

The two components of the motion can be analysed separately.

If air resistance is negligible, then there is zero force in the horizontal direction and this component of the velocity is constant.

The vertical component can be treated as a one-dimensional problem of an object in a gravitational field.

The path that the object follows is called a parabola.

WORKED EXAMPLE

A golf ball is hit so that it leaves the club at a velocity of 45 m s⁻¹ at an angle of 40° to the horizontal. Calculate:

a the horizontal component of the initial velocity

b the vertical component of the initial velocity

c the time taken for the ball to reach its maximum height

d the horizontal distance travelled when the ball is at its maximum height

(Ignore the effects of air resistance and spin on the ball.)

Answer

a $v_h = v\cos\theta = 45\cos 40 = 34.5\,\text{m}\,\text{s}^{-1}$

b $v_v = v\sin\theta = 45\sin 40 = 28.9\,\text{m}\,\text{s}^{-1}$

c In the vertical direction, the motion can be considered to be that of a ball thrown vertically upwards, decelerating under the effect of gravity. At the top of the flight, the vertical velocity will be, momentarily, zero. Use the equation:

$$v = u + at$$
$$0 = 28.9 + (-9.81)t$$
$$t = 2.94\,\text{m}\,\text{s}^{-1}$$

d The horizontal component of the velocity remains constant throughout the flight.

$$s = ut = 34.5 \times 2.94 = 101 = 100\,\text{m}$$

If the effects of air resistance and spin are ignored, the flight path would be symmetrical. This means that if the ball were hit on a level field, it would travel a total horizontal distance of 200 m before bouncing.

STUDY TIP

In the worked example, v_h, v_v and t are interim values used to find the value of s. Interim values should be calculated to one more significant figure than the original data, then the final answer for s is given to 2 significant figures.

8 A bullet of mass 50 g is fired horizontally from a height of 1.2 m. The bullet leaves the gun at a speed of 280 m s^{-1}.
  a Describe the path the bullet takes.
  b Assume the ground is level. Calculate:
    i the time that it takes for the bullet to hit the ground
    ii the distance the bullet travels before it hits the ground
  c State any assumptions you made in parts b i and b ii, and explain the effect they will have on your answer to part b ii.

> **END OF CHAPTER CHECK**

In this chapter, you have learnt to:
» define and use distance, displacement, speed, velocity and acceleration ☐
» use graphical methods to represent distance, displacement, speed, velocity and acceleration ☐
» determine velocity using the gradient of a displacement–time graph ☐
» determine acceleration using the gradient of a velocity–time graph ☐
» determine displacement from the area under a velocity–time graph ☐

» derive, from the definitions of velocity and acceleration, the equations of uniformly accelerated motion ☐
» solve problems using the equations of uniformly accelerated motion ☐
» solve problems involving the motion of objects falling in a uniform gravitational field without air resistance ☐
» describe an experiment to measure the acceleration of free fall using a falling object ☐
» describe and explain motion due to a uniform velocity in one direction and a uniform acceleration in a perpendicular direction ☐

3 Dynamics

Momentum and Newton's laws of motion

Concept of mass and weight

Before discussing Newton's laws of motion in detail, you need to revise the ideas of mass and weight and introduce the concept of momentum.

Mass and **weight** are often confused. Weight is the gravitational pull on an object and depends on the strength of the gravitational field at the position of the object.

In general, the two quantities are connected by the equation:

$W = mg$

where W is weight, m is mass and g is gravitational field strength (or acceleration of free fall).

>> The gravitational field strength near the surface of the Earth is $9.81\,N\,kg^{-1}$.
>> A mass of about $100\,g$ ($0.1\,kg$) has a weight of just less than $1\,N$ ($0.981\,N$) on the Earth's surface.
>> Its weight on the Moon is only $0.16\,N$ because the gravitational field strength on the Moon is only about ⅙ of that on Earth.

> **KEY TERMS**
>
> **Mass** is the property of an object that resists changes in motion. Mass is a base quantity and its unit, the kilogram, is a base unit.
>
> **Weight** is the gravitational pull on an object. Weight is a force and, like all forces, its unit is the newton (N).

Momentum

The concept of **momentum** is important in order to understand Newton's laws, which are discussed on pp. 28–30.

>> The unit of momentum is $kg\,m\,s^{-1}$.
>> It is calculated by multiplying a vector (velocity) by a scalar (mass) and is, therefore, a vector itself.
>> For example, an object of mass $2\,kg$ travelling at $3\,m\,s^{-1}$ has a momentum of $6\,kg\,m\,s^{-1}$.
>> An object of the same mass travelling at the same speed but in the opposite direction has a momentum of $-6\,kg\,m\,s^{-1}$.
>> It is important when you consider interactions between objects that you understand the vector nature of momentum.

> **KEY TERMS**
>
> **Momentum** (p) is defined as the product of mass and velocity:
>
> $p = mv$

> **STUDY TIP**
>
> The unit of momentum is, in base units, $kg\,m\,s^{-1}$. However, this is more usually referred to as $N\,s$.

WORKED EXAMPLE

Calculate the momentum of a cruise liner of mass $20\,000$ tonnes when it is travelling at $6.0\,m\,s^{-1}$.

(1 tonne = 1000 kg)

Answer

Convert the mass to kg:

$20\,000\,t = 20\,000 \times 1000\,kg = 20\,000\,000\,kg$

$p = mv = 20\,000\,000 \times 6.0 = 120\,000\,000\,kg\,m\,s^{-1} = 1.2 \times 10^8\,kg\,m\,s^{-1}$

Newton's first law　　　　　　　　　　REVISED ☐

> An object will remain at rest or move with constant velocity unless acted on by a resultant force.

» The first part of this law (an object will remain at rest unless acted upon by a resultant force) is relatively straightforward; we do not expect an object to move suddenly for no reason.
» The second part of the law (which refers to an object that will move with constant velocity unless it is acted upon by a resultant force) requires a little more thought.
» A golf ball putted along level ground will gradually slow down, as will a glider flying at a constant height in still air.
» In both these cases, frictional forces act in the opposite direction to the velocity of the object and cause it to decelerate.
» The frictional force on an object in a fluid is referred to as the viscous force as the object moves through the fluid.

When we observe motion on Earth, we cannot eliminate friction and we 'learn' (falsely) that a force is needed to keep objects moving. In practice, we only need that force to overcome frictional forces. If you think of a rock moving through outer space, there is no force on it – yet it will continue moving in a straight line forever, or until it encounters another object, perhaps in another galaxy.

Newton's second law　　　　　　　　　　REVISED ☐

> A resultant force acting on an object will cause a change in momentum in the direction of the force. The rate of change of momentum is proportional to the magnitude of the force.

KEY TERMS

Force is the rate of change of momentum.

Refer back to p. 27 to revise the meaning of momentum.

The first law describes what happens when there is *no* resultant **force** on an object. The second law explains what happens when there *is* a resultant force on the object. The second law defines force.

From this law, we can write:

$$F \propto \frac{\Delta p}{\Delta t}$$

The constant of proportionality defines the size of the unit of force. The newton is defined by making the constant equal to 1, when momentum is measured in $kg \, m \, s^{-1}$ and time is measured in s. Hence:

$$F = \frac{\Delta p}{\Delta t}$$

You see from this equation that force is measured in $kg \, m \, s^{-2}$.

$1 \, kg \, m \, s^{-2}$ is called 1 N (newton).

WORKED EXAMPLE

A golf ball of mass 45 g is putted along a level green with an initial velocity of $4.0\,\mathrm{m\,s^{-1}}$. It decelerates at a constant rate and comes to rest after $3.0\,\mathrm{s}$. Calculate the frictional force on the ball.

Answer

Convert the mass to kg:

$$45\,\mathrm{g} = \frac{45}{1000}\,\mathrm{kg} = 0.045\,\mathrm{kg}$$

initial momentum = $0.045 \times 4.0 = 0.18\,\mathrm{kg\,m\,s^{-1}}$

final momentum = 0

$$F = \frac{\Delta p}{\Delta t} = \frac{-0.18}{3.0} = -0.060\,\mathrm{N}$$

The minus sign in the answer shows that the force is acting in the opposite direction to the initial velocity.

Acceleration of a constant mass

In many situations, including the previous worked example, the mass of the object on which the resultant force is applied remains constant (or nearly constant). Consider the basic equation:

$$F = \frac{\Delta p}{\Delta t}$$

Now $\Delta p = \Delta(mv)$ and if m is constant this can be rewritten as $\Delta p = m\Delta v$. Therefore:

$$F = \frac{m\Delta v}{\Delta t}$$

But:

$$\frac{\Delta v}{\Delta t} = \text{acceleration}$$

So:

$$F = ma$$

The previous worked example could be solved using this equation, rather than using the rate of change of momentum.

It is important to recognise that the acceleration of an object is always in the same direction as the resultant force on the object.

The equation above also shows that the larger the mass of an object, the smaller the acceleration (or rate of change of velocity) of the object. Thus, mass can be seen to be the resistance to the change in motion of an object.

WORKED EXAMPLE

A car of mass 1.2 tonnes accelerates from $5\,\mathrm{m\,s^{-1}}$ to $30\,\mathrm{m\,s^{-1}}$ in $7.5\,\mathrm{s}$. Calculate the average accelerating force on the car.

Answer

$$\text{acceleration} = \frac{\text{change in velocity}}{\text{time taken}} = \frac{30 - 5}{7.5} = 3.3\,\mathrm{m\,s^{-2}}$$

Convert the mass to kilograms:

$$1.2\,\mathrm{t} = 1200\,\mathrm{kg}$$

force = mass × acceleration = $1200 \times 3.3 = 4000\,\mathrm{N}$

> ## NOW TEST YOURSELF
> TESTED ☐
>
> 3 A car of mass 1200 kg accelerates from rest to $18\,\mathrm{m\,s^{-1}}$ in $6.3\,\mathrm{s}$. Calculate:
>   a the acceleration of the car
>   b the average resultant force acting on the car
>   c the momentum of the car when it is travelling at $18\,\mathrm{m\,s^{-1}}$

Newton's third law

The third law looks at the interaction between two objects.

> If an object A exerts a force on an object B, then object B will exert a force on object A of equal magnitude but in the opposite direction.

(a) **(b)** **(c)**

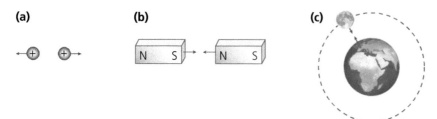

▲ Figure 3.1 (a) Two protons repel each other, (b) two magnets attract each other, (c) the Earth and the Moon attract each other

The examples in Figure 3.1 show forces on two objects of roughly similar size. It is easy to appreciate that the forces in each example are of equal size.

However, it is also true with objects of very different sizes. For example, when you jump off a wall (as in Figure 3.2), there is a gravitational pull on you from the Earth that pulls you down towards the ground. What you do not think about is that you also pull the Earth upwards towards you with an equal sized force. Of course, the movement of the Earth is negligible because it is so much more massive than you are – but the force is still there.

The child is pulled down by the Earth with a force, W

The Earth is pulled up by the child with a force, W

▲ Figure 3.2 Interaction between two objects

> **NOW TEST YOURSELF** TESTED ☐
>
> 4 In Figure 3.1, in what direction does the Moon accelerate?
> 5 In what direction does a ball thrown vertically upwards from the Earth's surface accelerate?

Non-uniform motion

Effect of air resistance

Air resistance, sometimes called drag, affects all moving objects near the Earth's surface, including the motion of falling objects. It works like this:

» Air resistance depends on the shape of an object.
» The resistance on a streamlined object is less than on a non-streamlined object.
» Air resistance also depends on the speed at which the object travels.
» Air resistance increases as the speed of the object increases.
» The drag force on a falling object increases as the object accelerates.
» The resultant force (= weight – drag force) decreases as the speed increases.
» Therefore, the acceleration decreases.

When the drag force is equal to the gravitational pull on the object, it will no longer accelerate but will fall with a constant velocity. This velocity is called the **terminal velocity**.

Figure 3.3 shows how the speeds of a shuttlecock and of a tennis ball change as they fall from rest.

STUDY TIP

An object falls with its **terminal velocity** when the total upward forces on the object, as it falls through a uniform gravitational field, are equal to the object's weight.

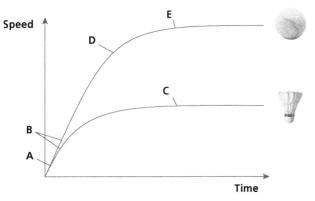

▲ Figure 3.3

In Figure 3.3:

» at **point A**, the air resistance (or drag force) is negligible and both the shuttlecock and the tennis ball fall with the same acceleration, g
» at **point B**, the air resistance (compared with the weight of the ball) remains small and it continues to fall with the same acceleration; the shuttlecock has a much smaller weight than the ball and the air resistance on it is significant compared with its weight, so its acceleration is reduced
» at **point C**, the air resistance is equal to the weight of the shuttlecock. It no longer accelerates and falls with its terminal velocity
» at **point D**, the air resistance on the ball is now significant and its acceleration is reduced
» at **point E**, the air resistance is equal to the weight of the ball and it falls with its terminal velocity

NOW TEST YOURSELF

TESTED ☐

6 A ball-bearing falls at a constant speed through oil. Name the forces acting on it in the vertical direction and state the magnitude of the resultant force on it.

Linear momentum and its conservation

Principle of conservation of momentum

REVISED ☐

» Newton's third law leads to the conclusion that in any interaction momentum is conserved.
» This means that the total momentum of a closed system (that is, a system on which no external forces act) is the same after an interaction as before the interaction.

Consider two objects that move towards each other, as in Figure 3.4, and then stick to each other after the collision.

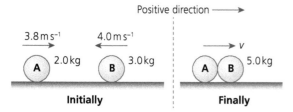

▲ **Figure 3.4 Collision between two objects**

If we consider the positive direction to be from left to right:

total momentum before the collision = total momentum after the collision

$$(2.0 \times 3.8) + (3.0 \times -4.0) = 5v$$

$$-4.4 = 5v$$

$$v = -0.88\,\text{m s}^{-1}$$

The negative sign means that the velocity after the collision is from right to left.

A formal statement of the law is as follows:

> The law of conservation of momentum states that the total momentum of a closed system before an interaction is equal to the total momentum of that system after the interaction.

NOW TEST YOURSELF

TESTED ☐

7 A trolley of mass 250 g travelling at 2.4 m s⁻¹ collides with and sticks to a second stationary trolley of mass 400 g. Calculate the speed of the trollies after the impact.

Collisions in two dimensions

REVISED ☐

The previous example considers a head-on collision, where all the movement is in a single direction.

The law applies equally if there is a glancing collision and the two objects move off in different directions. In this type of problem, the individual momenta must be resolved so that the conservation of momentum can be considered in two perpendicular directions.

Before collision   After collision

▲ **Figure 3.5**

Figure 3.5 shows a disc A of mass m_A moving towards a stationary disc B of mass m_B with a velocity u. The discs collide. After the collision, disc A moves off with velocity v_A at an angle θ to its original velocity and disc B moves with a velocity v_B at an angle of ϕ to the original velocity of A.

Momenta parallel to u:

momentum before collision = $m_A u$

momentum after collision = $m_A v_A \cos\theta + m_B v_B \cos\phi$

Check your answers at **www.hoddereducation.com/cambridgeextras**

Therefore:

$$m_A u = m_A v_A \cos\theta + m_B v_B \cos\phi$$

Momenta perpendicular to u:

momentum before collision = 0

momentum after collision = $m_A v_A \sin\theta + m_B v_B \sin\phi$

Therefore:

$$0 = m_A v_A \sin\theta + m_B v_B \sin\phi$$

WORKED EXAMPLE

A particle moves towards a stationary particle of equal mass with a velocity of $2.00\,\text{m s}^{-1}$. After the collision, one particle moves off with a velocity $1.00\,\text{m s}^{-1}$ at an angle of 60° to the original velocity. The second particle moves off with a velocity of $1.73\,\text{m s}^{-1}$. Calculate the angle the second particle makes with the original velocity.

Answer

Momenta perpendicular to $u = 0$

Vertical momentum after collision = $1.00m\sin 60 + 1.73m\sin\theta$

Therefore:

$$1.00m\sin 60 = -1.73\,m\sin\theta$$

$$\frac{0.866}{1.73} = -\sin\theta$$

$$\theta = -30°$$

Types of interactions

Elastic interactions

» In an elastic interaction, not only is the momentum conserved, but the kinetic energy is also conserved.
» On the macroscopic scale, this is rare. However, many interactions do approximate to being perfectly elastic and the mathematics of an elastic interaction can be used to model these.
» On the microscopic scale, for example, the collision between two charged particles, such as protons, can be considered to be elastic.

It is worth noting that in any perfectly elastic collision, the relative speed of approach is equal to the relative speed of separation after the interaction. A good example of this is the nearly elastic interaction of a golf ball being struck by the much more massive club (Figure 3.6).

> **STUDY TIP**
>
> A glancing elastic collision between two equal masses always leads to the two masses having velocities after the collision that are perpendicular to each other. Refer back to the worked example above, which involves an elastic collision.

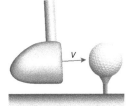

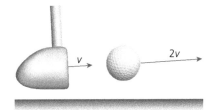

The golf club approaches the ball at a velocity of v.

The club continues to move at a velocity of (very nearly) v. The ball moves off at a speed of (nearly) $2v$. The speed of separation of the ball from the club is equal to the speed of approach of the club to the ball.

▲ **Figure 3.6 A (nearly) perfectly elastic collision**

Inelastic interactions

» In an inelastic interaction, some of the initial kinetic energy is converted into other forms, such as sound and internal energy.

» In an inelastic interaction the total kinetic energy before the interaction is not equal to the total kinetic energy after the interaction. In some collisions the total kinetic energy after the collision is less than before the collision – the total kinetic energy decreases, whereas in an explosion there is zero kinetic energy prior to the explosion, but plenty of kinetic energy afterwards – the total kinetic energy increases.

» As in *all* interactions, momentum is conserved.

There are numerous examples and degrees of inelastic collision – from nearly perfectly elastic, such as one billiard ball striking another, to two objects sticking together, such as two identical trolleys colliding and moving off together, as shown in Figure 3.7.

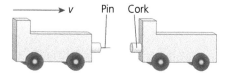

One trolley moves towards a second identical stationary trolley with a speed v.

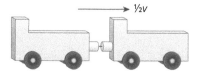

The two trolleys stick together and move off with a combined speed of $\frac{1}{2}v$.

▲ **Figure 3.7 An inelastic collision**

WORKED EXAMPLE

A glider of mass 0.20 kg on an air track is moving at 3.6 m s^{-1} towards a second glider of mass 0.25 kg, which is moving at 2.0 m s^{-1} in the opposite direction. When the two gliders collide, they stick together.

a Calculate their joint velocity after the collision.

b Show that the collision is inelastic.

Answer

a momentum before the collision = $(0.20 \times 3.6) + (0.25 \times -2.0) = 0.22$ kg m s^{-1}

momentum after the collision = $(0.20 + 0.25)v = 0.45v$, where v is the velocity of the two gliders after the collision

momentum after the collision = momentum before the collision

$$0.22 = 0.45\,v$$

$$v = 0.49\,\text{m s}^{-1}$$

b kinetic energy before the collision = $(\frac{1}{2} \times 0.2 \times 3.6^2) + (\frac{1}{2} \times 0.25 \times 2.0^2)$

$$= 1.3 + 0.5$$

$$= 1.8\,\text{J}$$

kinetic energy after the collision = $(\frac{1}{2} \times 0.45 \times 0.49^2)$

$$= 0.054\,\text{J}$$

The kinetic energy after the collision is less than the kinetic energy before the collision, therefore the collision is inelastic.

> **STUDY TIP**
>
> Note that each step in the answer is clearly explained and that the final comment completes the answer.

> ## NOW TEST YOURSELF

8   A ball of mass 250 g travelling at 13 m s^{-1} collides with and sticks to a second stationary ball of mass 400 g.

    a   Calculate the speed of the balls after the impact.

    b   Show whether or not the collision is elastic.

9   A disc of mass 2.4 kg is moving at a velocity of 6.0 m s^{-1} at an angle of 40° west of north. Calculate its momentum in:

    a   the westerly direction

    b   the northerly direction

PRACTICAL SKILL

Masses are often supplied with their mass written on them: 10 g, 50 g, 100 g, etc. Do not assume that these values are accurate. Check the mass using a balance if possible.

> ## END OF CHAPTER CHECK

In this chapter, you have learnt to:

» understand that mass is the property of an object which resists change in motion ▢

» define and use linear momentum as *mass × velocity* ▢

» describe weight as the effect of a gravitational field on a mass ▢

» recall and use the weight of an object is equal to the product of its mass and the acceleration of free fall, *g* ▢

» understand that the acceleration of an object is in the same direction as the resultant force on the object ▢

» define and use force as the *rate of change of momentum* ▢

» recall and use the equation $F = ma$ ▢

» state and apply Newton's three laws of motion ▢

» have a qualitative understanding of frictional forces and viscous forces ▢

» describe and explain the motion of objects in a uniform gravitational field with air resistance ▢

» understand that objects moving against frictional forces may reach a terminal velocity ▢

» state the principle of conservation of momentum and understand that in *all* interactions momentum is conserved ▢

» apply the principle of conservation of momentum to simple problems ▢

» apply the principle of conservation of momentum to problems in two dimensions ▢

» understand that in a perfectly elastic interaction the total kinetic energy is conserved ▢

» recall and use that in a perfectly elastic interaction the relative speed of approach is equal to the relative speed of separation ▢

» understand that in an inelastic interaction there will be changes in the total kinetic energy ▢

» apply the principle of conservation of momentum to both elastic and inelastic interactions ▢

Forces, density and pressure

Turning effects of forces

Centre of gravity

Centre of gravity is a term that is often used in everyday life.

»» When we use the term 'centre of gravity', we are implying that all the weight of the object is concentrated at this single point. This is not the case – the weight of an object does not act from a single point but is spread through all the particles of the object.
»» However, it is often convenient to consider the weight acting at a single point – this point is called the **centre of gravity** of the object.
»» An object with a low centre of gravity is more stable and is less likely to be knocked over than an object with a higher centre of gravity.

> **KEY TERMS**
>
> The **centre of gravity** of an object is the point at which its whole weight may be considered to act.
>
> A **uniform object** has its centre of gravity at the geometric centre of the object.

Moment of a force

The turning effect of a force about a point (sometimes known as the torque) is known as its **moment** about that point. When considering a single force, the point about which the force is producing its turning effect must be specified.

The moment of a force about a point = the magnitude of the force × the perpendicular distance from the force to the point

In symbols, this equation is:

$T = F \times d$

The SI unit of the moment of a force about a point is N m.

Figure 4.1 shows a spanner turning a nut.

»» The force is not perpendicular to the spanner.
»» Therefore, either the component of the force perpendicular to the spanner or the perpendicular distance from the centre of the nut to the line of action of the force must be used in the calculation.
»» The perpendicular distance of the line of action of the 30 N force from the centre of the nut is the distance x.

$x = 0.25 \cos 20 = 0.235\,\text{m}$

»» Hence, the moment about the centre of the nut is:

$30 \times 0.235 = 7.05\,\text{N m} \approx 7\,\text{N m}$

> **KEY TERMS**
>
> The **moment** of a force about a point equals the force multiplied by the perpendicular distance of the line of action of the force from the point.

> **STUDY TIP**
>
> A force does not have a unique moment. The moment depends on the point about which the force has a turning effect. Therefore, you should always refer to the point about which the moment is produced.

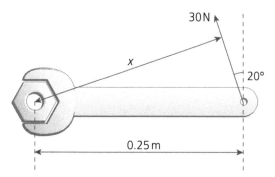

▲ **Figure 4.1 Turning forces on a spanner**

Check your answers at **www.hoddereducation.com/cambridgeextras**

NOW TEST YOURSELF

1 Calculate the moment that is produced about the nut in Figure 4.1 when the 30 N force is applied:
   a along the dashed line (i.e. perpendicular to the spanner shaft)
   b at an angle of 60° to the dashed line
2 The perpendicular distance from a door handle to the hinges is 0.66 m. A force of 4.0 N perpendicular to the door is applied to the handle. Calculate the torque produced about the hinges.

> **STUDY TIP**
> Torque is a vector. If a torque that tends to turn an object in a clockwise sense is considered to be positive, then a torque that tends to cause the object to move in an anticlockwise sense is considered negative.

Couple

» A couple is produced when two parallel forces of equal magnitude act in opposite directions and have lines of action that do not coincide.
» For example, you apply a couple when you turn on a tap.
» When considering the **torque of a couple**, you do not need to state the specific point that the torque is produced about.
» A couple produces rotation only.
» A single force may produce rotation, but it will always tend to produce acceleration as well (Figure 4.2).

> **KEY TERMS**
> **Torque of a couple** is the magnitude of one of the forces × the perpendicular distance between the lines of action of the forces.

(a)  **(b)**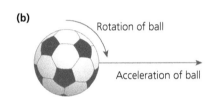

▲ Figure 4.2 (a) A footballer kicks the ball, striking the side of the ball. (b) The ball accelerates but also tends to rotate in a clockwise direction

NOW TEST YOURSELF

3 The handle of a screwdriver has a diameter of 1.8 cm. Forces each of magnitude 4.2 N act in opposite directions on opposite sides of the handle. Calculate the torque produced by the couple.

4 Two forces of magnitude 8.0 N act in opposite directions on either side of a bolt head of diameter 2.4 cm. Calculate the torque produced by the couple.

Equilibrium of forces

Equilibrium

A point object is in equilibrium when the resultant force acting on the object is zero.

For an object of finite size to be in equilibrium:

» the resultant force on the object must be zero
» the resultant torque on the object must be zero

Principle of moments

The principle of moments is a restatement of the second condition for an object to be in equilibrium:

> For an object to be in equilibrium, the sum of the moments about any point is zero.

A useful way of using this when you are considering coplanar forces is to say 'the clockwise moments equal the anticlockwise moments'.

WORKED EXAMPLE

A student has a uniform metre rule of weight 1.20 N. He attaches a weight of 1.50 N at the 10.0 cm point and places the ruler on a pivot. He adjusts the position of the pivot until the ruler balances. Deduce the position of the pivot.

Answer

Draw a diagram of the set-up (Figure 4.3).

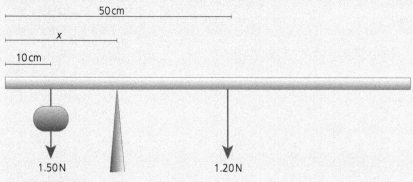

▲ **Figure 4.3**

The ruler is uniform and therefore the centre of gravity is at its centre.

Take moments about the pivot:

    clockwise moment $= (50 - x) \times 1.20 \,\text{N cm}$

    anticlockwise moment $= (x - 10) \times 1.50 \,\text{N cm}$

For equilibrium, the clockwise moments = the anticlockwise moments:

    $(50 - x) \times 1.2 = (x - 10) \times 1.5$

                 $x = 27.8 \,\text{cm}$

PRACTICAL SKILL

Practicals using moments involve measuring lengths with a metre rule. The smallest scale division on a metre rule is 0.1 cm, so ensure all lengths are recorded to 0.1 cm. Even if the value is 22.0 cm, the zero must be included in the recorded data. For further information on recording data see 'Presentation of data and observations' on p. 102.

▶ NOW TEST YOURSELF

TESTED ☐

5 A uniform metre rule is pivoted on the 30 cm mark. When a weight of 2.0 N is hung from the 14 cm mark, the ruler balances. Calculate the mass of the ruler.

Check your answers at **www.hoddereducation.com/cambridgeextras**

Representing forces in equilibrium

In the section on vectors, you met the idea of using vector diagrams to add vectors acting at different angles, and you also met the concept of resolving vectors into their component parts. Vector diagrams can also be used when an object is in equilibrium.

A lamp of weight W is suspended from the ceiling by a cord. The lamp is pulled to one side with a horizontal force F so that the cord makes an angle θ with the vertical, as shown in Figure 4.4

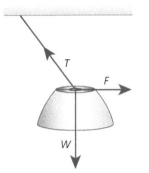

W

The diagram shows the forces acting on the lamp.
T is the tension in the flex.

Draw the vectors
W and F.

Complete the triangle.
The third side represents
the tension T.

▲ **Figure 4.4 Forces in equilibrium**

Note the difference between this and using the vector triangle to find the resultant of two forces. This is, again, a closed triangle, but with all the arrows going in the same direction round the triangle. This shows that the sum of these three forces is zero and that the object is in equilibrium.

WORKED EXAMPLE

A helium balloon is tethered to the ground using a cable that can withstand a maximum force of 10.0 kN before breaking. The net upward force on the balloon due to its buoyancy is 8.0 kN (Figure 4.5).

Calculate the maximum horizontal force the wind can produce on the balloon before the cable snaps, and the angle the cable makes with the vertical when this force is applied.

Answer
Refer to Figure 4.6:
1 Draw a vertical arrow of length 4.0 cm to represent the upthrust on the balloon.
2 Draw a horizontal construction line from the top of the vertical arrow.
3 Separate the needle and the pencil tip of your compasses by a distance of 5 cm to represent the tension force in the cable. Place the needle on the bottom of the vertical line and draw an arc to intersect with the horizontal construction line.
4 Join from the intersection to the bottom of the vertical line. This represents the tension in the cable.
5 Draw in the arrow to represent the horizontal force from the wind.

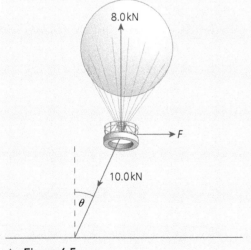

▲ **Figure 4.5**

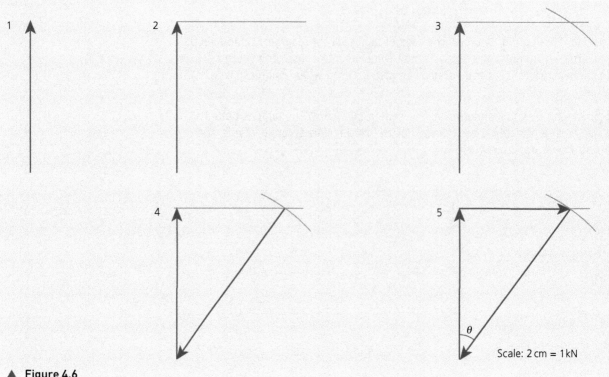

▲ Figure 4.6

The length of the horizontal arrow = 3.0 cm.
Therefore, the force due to the wind is 6.0 kN.

The angle with the vertical, θ, measured with a protractor = 37°.

Scale: 2 cm = 1 kN

> **NOW TEST YOURSELF**   TESTED ☐

6 A uniform picture of weight 3.6 N is attached to a wall using a string, as shown in Figure 4.7. Each end of the string makes an angle of 40° with the horizontal. Calculate the tension in the string.

▲ Figure 4.7

7 A uniform metre rule is pivoted at its midpoint. A weight of 6.00 N is hung from the 30.0 cm mark. The ruler is held in equilibrium by a string attached to the 10 cm mark making an angle of 60° with the ruler. Calculate the tension in the string.

Density and pressure

Density

You will have met the concept of **density** in earlier work.

The unit of density is kilogram per metre cubed ($kg\,m^{-3}$) or gram per centimetre cubed ($g\,cm^{-3}$).

WORKED EXAMPLE

A beaker has a mass of 48 g. When 120 cm^3 of copper sulfate solution are poured into the beaker, it is found to have a total mass of 174 g. Calculate the density of the copper sulfate solution.

Answer

mass of copper sulfate solution = $174 - 48 = 126\,g$

$$\rho = \frac{m}{V} = \frac{126}{120} = 1.05\,g\,cm^{-3}$$

> **KEY TERMS**
>
> The **density** (ρ) of a substance is defined as the mass per unit volume of the substance:
>
> density, $\rho = \dfrac{mass}{volume}$

> ### NOW TEST YOURSELF
>
> TESTED
>
> 8 A beaker contains 320 g of a liquid of density $1.2 \times 10^3\,kg\,m^{-3}$. Calculate the volume of this liquid.
>
> 9 A cube of iron has sides of length 8.0 cm. Given the density of iron is $7.3\,g\,cm^{-3}$, calculate the mass of the cube.

Pressure

Pressure can easily be confused with force, the difference being that pressure considers the area on which the force acts.

Pressure is measured in newtons per metre squared ($N\,m^{-2}$). $1\,N\,m^{-2}$ is called 1 **pascal (Pa)**. It is sometimes convenient to use $N\,cm^{-2}$.

Pressure, unlike force, is a scalar. Therefore, pressure does not have a specific direction.

> **KEY TERMS**
>
> **Pressure** (p) is defined as the normal force per unit area:
>
> pressure, $p = \dfrac{force}{area}$
>
> 1 **pascal** is the pressure exerted by a force of 1 newton acting normally on an area of 1 metre squared.

WORKED EXAMPLE

Coins are produced by stamping blank discs with a die. The diameter of a blank disc is 2.2 cm and the pressure on the disc during stamping is $2.8 \times 10^5\,Pa$. Calculate the force required to push the die against the blank disc.

Answer

$$\text{area of the coin} = \pi\left(\frac{d}{2}\right)^2 = \pi\left(\frac{2.2}{2}\right)^2 = 3.8\,cm^2 = 3.8 \times 10^{-4}\,m^2$$

$$\text{pressure} = \frac{force}{area}$$

Hence:

$$\text{force} = \text{pressure} \times \text{area} = 2.8 \times 10^5 \times 3.8 \times 10^{-4} = 106\,N$$

Pressure in a liquid

A liquid exerts pressure on the sides of its container and on any object in the liquid. The pressure exerted by the liquid increases as the depth increases.

Figure 4.8 shows a beaker containing a liquid of density ρ. To calculate the pressure on the area A due to the weight of the column of water of height Δh above it:

weight = mass × g (where g is the gravitational field strength)

mass of the column = density × volume, where the volume of the column of water = $A \times \Delta h$

mass of the column = $\rho \times A \times \Delta h$

weight of the column = $\rho \times A \times \Delta h \times g$

pressure on area $A = \dfrac{\text{force}}{\text{area}} = \dfrac{\text{weight}}{\text{area}}$

$$= \frac{\rho \times A \times \Delta h \times g}{A}$$

pressure = $\rho \Delta h g$

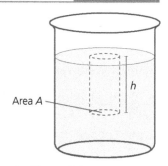

▲ Figure 4.8

WORKED EXAMPLE

Atmospheric pressure is 1.06×10^5 Pa. A diver descends to a depth of 24 m in seawater of density 1.03×10^3 kg m^{-3}. Calculate the total pressure on the diver.

Answer

pressure due to seawater = $\rho \Delta h g = (1.03 \times 10^3) \times 24 \times 9.81 = 2.425 \times 10^5$ Pa

total pressure = $(2.425 \times 10^5) + (1.06 \times 10^5) = 3.485 \times 10^5 = 3.49 \times 10^5$ Pa

STUDY TIP
The total pressure is equal to the pressure due to the water plus atmospheric pressure. It is easy to forget to include atmospheric pressure.

NOW TEST YOURSELF

10 Calculate the pressure the iron cube in question 9 exerts on a bench as it rests on one of its faces.

11 Oil of density 850 kg m^{-3} is poured into a measuring cylinder to a depth of 0.800 m. Calculate the pressure exerted on the base of the measuring cylinder by the oil.

12 A simple mercury barometer has a column of height 760 mm. Calculate the pressure the mercury exerts at the bottom of this column. (density of mercury = 13 600 kg m^{-3})

Upthrust

» Upthrust (or the buoyancy force) in a fluid is caused by the increase in hydrostatic pressure as the depth of the fluid increases.
» Consider a rectangular box in a liquid – the bottom of the box is at a greater depth than the top.
» Thus, the pressure on the bottom is greater than the pressure on the top.
» As pressure = force/area and since the two surfaces have the same area, the force on the bottom is greater than the force on the top and the box is pushed upwards.

Archimedes' principle

Archimedes' principle states that the upthrust on an object in a fluid is equal to the weight of fluid displaced.

This follows on from the explanation above.

Check your answers at **www.hoddereducation.com/cambridgeextras**

WORKED EXAMPLE

An object of mass 15.0 kg and density 1200 kg m⁻³ is fully immersed in a liquid of density 1050 kg m⁻³.
Calculate:

a the upthrust on the object
b the resultant force on the object

(Take $g = 9.81\,N\,kg^{-1}$)

Answer

a The volume of the object is calculated from the formula $\rho = m/V$:

$$V = \frac{m}{\rho} = \frac{15.0}{1200} = 0.0125\,m^3$$

$F = \rho g V$ where ρ is the density of the liquid

$$= 1050 \times 9.81 \times 0.0125$$

$$= 129\,N$$

b resultant force $= mg +$ upthrust

$$= (15.0 \times 9.81) - 129$$

$$= 18\,N$$

The minus sign before 129 arises because the upthrust is in the opposite direction to the weight.

> ## NOW TEST YOURSELF
> TESTED ☐
>
> 13 Consider a cuboid of height y and cross-sectional area A. The cuboid is immersed in a liquid of density ρ so that its upper surface is at a depth h below the surface of the liquid.
>   a Determine the weight of the liquid the cuboid displaces.
>   b Determine the pressure on the upper and lower surfaces of the cuboid.
>   c Determine the downward force on the upper surface of the cuboid due to the pressure.
>   d Determine the upward force on the lower surface of the cuboid due to the pressure.
>   e Determine the difference between these two forces.
>   f Compare your answers to parts **a** and **e** and comment.

Question 13f leads you to the equation linked with Archimedes' principle:

$$F = \rho g V$$

where V is the volume of the object.

> ## REVISION ACTIVITY
>
> 'Must learn' equations:
>
> density $(\rho) = \dfrac{mass}{volume}$ $\qquad$ pressure $(p) = \dfrac{force}{area}$

> ## END OF CHAPTER CHECK
>
> In this chapter, you have learnt:
> » to understand that the centre of gravity of an object is the point at which we may consider the whole weight of the object to act ☐
> » to define and apply the moment of a force about a point ☐
> » to understand that a couple is a pair of forces which act to produce rotation only ☐
> » to define and apply the torque of a couple ☐
> » to state and apply the principle of moments ☐
> » that a system is in equilibrium when the resultant force and the resultant torque on the system are both zero ☐
> » to use the vector triangle to represent coplanar forces in equilibrium ☐
> » to define and use the concept of density ☐
> » to define and use the concept of pressure ☐
> » to understand that the upthrust force is caused by a difference in the hydrostatic pressure ☐
> » to calculate the upthrust on an object in a fluid using Archimedes' principle ($F = \rho g V$) ☐

5 Work, energy and power

Work and efficiency

Work

Work done has a precise meaning in physics and care must be taken when using this term. The unit of work is the **joule** (J).

Work = force × displacement in the direction of the line of action of the force.

Both force and displacement are vectors. Note that for work to be done there must be a component of the force that is parallel to the displacement.

> **KEY TERMS**
>
> **Work** is defined as being done when a force moves its point of application in the direction in which the force acts.
>
> **1 joule** of work is done when a force of 1 newton moves its point of application 1 metre in the direction of the force.

Calculating work done

When calculating work done, care must be taken that the force and the displacement are parallel. Consider a child sliding down a slide (Figure 5.1).

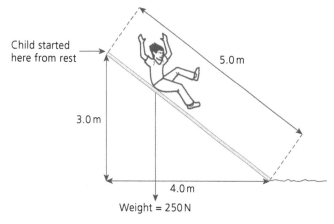

Child started here from rest

5.0 m

3.0 m

4.0 m

Weight = 250 N

▲ **Figure 5.1**

The force on the child is the child's weight, 250 N, which acts vertically downwards. The total distance moved is 5.0 m, but the displacement *parallel to the force* is only 3.0 m. So:

work done = 250 N × 3.0 m = 750 J

It is worth noting that, in this example, the work is done on the child by gravity, rather than the child doing the work.

In general:

work done = component of the force parallel to displacement × the displacement

Look at Figure 5.2. By resolving the force F into vertical and horizontal components, you can see that the component parallel to the displacement is $F\cos\theta$. Therefore:

work done $= Fx\cos\theta$

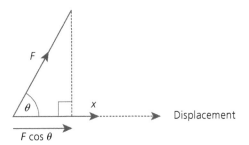

▲ Figure 5.2

WORKED EXAMPLE

Figure 5.3 shows a man wheeling a barrow.

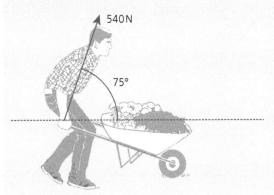

▲ Figure 5.3

He applies a force of 540 N to the barrow in a direction 75° from the horizontal. He moves the barrow 30 m along the level ground. Calculate the work he does against friction.

Answer

work done $= Fx\cos\theta$

work done $= 540 \times 30 \times \cos 75 = 4200\,\text{J}$

> **NOW TEST YOURSELF**   TESTED ☐

1 Calculate the work done when a force of magnitude 4.5 N moves its point of application 6.0 m:
   a   in the direction of the line of action of the force
   b   at 30° to the line of action of the force
   c   at 90° to the line of action of the force
2 An incline is at an angle of 30° to the horizontal. A force of 25 N pulls a box 4.0 m along the incline. Calculate:
   a   the total work done by the force
   b   the work done against gravity by the force

Energy conversion and conservation

Energy   REVISED ☐

Energy is not an easy concept and, like work, it has a precise meaning in physics. Also, like work, energy is measured in joules. When an object has 300 J of energy, it means that it can do 300 J of work. Different forms of energy are shown in Table 5.1.

> **KEY TERMS**
> **Energy** is defined as the ability (or capacity) to do work.

Type of energy	Description
Kinetic	The ability to do work due to the movement of an object
Gravitational potential	The ability to do work due to the position of an object in a gravitational field
Elastic potential	The ability to do work due to the deformation of an object (e.g. a compressed or extended spring)
Sound	The ability to do work due to the kinetic and potential energy of the vibrating particles in a sound wave
Internal	The sum of the random kinetic and potential energies of the molecules in an object
Electric potential	The ability to do work due to the position of a charged particle in an electric field
Chemical potential	The ability to do work due to potential energy of the particles making up substances
Nuclear potential	The ability to do work due to the potential energy of the subatomic particles in the nuclei of atoms

Energy conversion and efficiency

REVISED ☐

Machines are used to do work, converting energy from one form to another. In practice, machines are never 100% efficient. This means that the total energy input is greater than the useful work output. Some of the energy input is converted to unwanted forms such as thermal energy.

$$\text{efficiency of a machine} = \frac{\text{useful energy output}}{\text{total energy input}} \times 100\%$$

Efficiency is quoted either as a ratio or a percentage. Consequently, efficiency has no units.

WORKED EXAMPLE

A petrol motor is used to lift a bag of sand of mass 2700 kg from the ground up to a window 12 m above the ground. Eighteen per cent of the input energy is converted into gravitational potential energy of the sand.

a Calculate the energy input to the motor.

b Discuss the energy changes involved in the process.

Answer

a $\text{efficiency of a machine} = \frac{\text{useful energy output}}{\text{total energy input}} \times 100\%$

useful work done $= mgh = 2700 \times 9.81 \times 12 = 317\,844\,\text{J}$

$18 = \frac{317\,844}{\text{energy input}} \times 100\%$

$\text{energy input} = \frac{317\,844 \times 100}{18} = 1\,765\,800\,\text{J} \approx 1.8\,\text{MJ}$

b The chemical potential energy of the petrol and oxygen is transferred to gravitational potential energy and thermal energy of the surroundings. Eighteen percent of the energy is used to do work against gravity in lifting the bag of sand. The remainder is transferred to the surroundings as thermal energy.

STUDY TIP

What happens inside the motor is complex. There are transient forms of energy, such as elastic potential energy, as the fuel is burnt and put under pressure, and the conversion of this to kinetic energy of the oscillating piston. This is not needed for the exam and will not get you any marks.

Conservation of energy

REVISED ☐

The law of **conservation of energy** is often expressed as:

» Energy cannot be created or destroyed. It can only be converted from one form to another.

» The total energy of a closed system will be the same before an interaction as after it.

KEY TERMS

The law of **conservation of energy** states that the total energy of a closed system is constant.

Check your answers at **www.hoddereducation.com/cambridgeextras**

When energy is transformed from one form to another either:

» work is done – for example, a man does work against gravity by lifting a large mass onto his shoulders

or

» energy is radiated or received in the form of electromagnetic radiation – for example, thermal energy is radiated away from the white-hot filament of a lamp by infrared and light radiation.

> ## NOW TEST YOURSELF
> TESTED ☐
>
> 3 A crane on a building site lifts a pallet of bricks of mass 3.5×10^3 kg from the ground to a height of 22 m.
>   a Calculate the work done against gravity.
>   b The input energy to the crane motor is 2.5 MJ. Calculate the efficiency of the system.
> 4 The efficiency of a diesel motor is 35%. Calculate how much useful work is done by the motor for each megajoule of input energy.

Potential energy and kinetic energy

Gravitational potential energy
REVISED ☐

Consider a mass m lifted through a height Δh.

» The weight of the mass is mg, where g is the gravitational field strength.

work done = force × distance moved

$$= mg\Delta h$$

» Due to its new position, the object is now able to do extra work equal to $mg\Delta h$.
» It has gained extra **gravitational potential energy**, $\Delta E_p = mg\Delta h$:

change in gravitational potential energy $= mg\Delta h$

In these examples, we have considered objects close to Earth's surface, where we can consider the gravitational field to be uniform. In your A Level studies, you will explore this further and consider examples where the gravitational field is not uniform.

Kinetic energy
REVISED ☐

Consider an object of mass m, at rest, which accelerates to a speed of v over a distance s.

work done in accelerating the object = force × distance

$$W = Fs$$

But:

$$F = ma$$

In the equation $v^2 = u^2 + 2as$, the object starts at rest so that means $u = 0$. Hence:

$$F = ma = \frac{mv^2}{2s}$$

$$W = Fs = \frac{mv^2}{2s}s = \tfrac{1}{2}mv^2$$

The object is now able to do extra work $= \tfrac{1}{2}mv^2$ due to its speed. It has **kinetic energy** $E_k = \tfrac{1}{2}mv^2$.

WORKED EXAMPLE

A cricketer bowls a ball of mass 160 g at a speed of 120 km h^{-1}. Calculate the kinetic energy of the ball.

Answer

Convert the speed from km h^{-1} to m s^{-1}:

$$120 \text{ km h}^{-1} = 120 \times \frac{1000}{3600} \text{ m s}^{-1} = 33.3 \text{ m s}^{-1}$$

Convert 160 g to kg = 0.16 kg.

$$E_k = \tfrac{1}{2}mv^2 = \tfrac{1}{2} \times 0.16 \times 33.3^2$$

$$E_k = 89 \text{ J}$$

NOW TEST YOURSELF

5 Calculate the change in gravitational potential energy when a rock of mass 2.4 kg is raised through a height of 30 m.

6 Calculate the kinetic energy of a tennis ball of mass 57 g travelling at a speed of 40 m s^{-1}.

7 A ball of mass 0.30 kg initially at rest falls from a height of 25 m. It hits the ground at a speed of 22 m s^{-1}. (The effect of air resistance is negligible.) Calculate:

   a the kinetic energy gained by the ball

   b the gravitational potential energy lost by the ball

PRACTICAL SKILL

When investigating energy changes involving the expression $\Delta E_p = mg\Delta h$, remember that Δh is the vertical change in the height. Practical investigations often require a vertical height to be measured. A set square should be used to ensure the height being measured is vertical.

Power

Power (P) is the rate of doing work or transforming energy.

The unit of power is the **watt (W)**.

There is a power of 1 watt when energy is transferred or work is done at the rate of 1 joule per second.

KEY TERMS

$$\text{power} = \frac{\text{work done}}{\text{time taken}}$$

$$= \frac{\text{energy transformed}}{\text{time taken}}$$

In symbols:

$$P = \frac{W}{t}$$

WORKED EXAMPLE

A pebble of mass 120 g is fired from a catapult. The pebble accelerates from rest to 15 m s^{-1} in 0.14 s. Calculate the average power input to the pebble during the firing process.

Answer

$120 \text{ g} = 0.12 \text{ kg}$

gain in kinetic energy $= \tfrac{1}{2}mv^2 = 0.5 \times 0.12 \times 15^2 = 13.5 \text{ J}$

$$\text{power} = \frac{13.5}{0.14} = 96 \text{ W}$$

Check your answers at **www.hoddereducation.com/cambridgeextras**

Power and velocity

Consider a car travelling at a constant velocity v along a straight, level road. The engine must continue to do work against friction. If the frictional force is F, then the engine will supply an equal-sized force in the opposite direction. The work done by the engine, ΔW, in time Δt is $F\Delta x$, where Δx is the distance travelled in time Δt:

$$\text{power} = \frac{F\Delta x}{\Delta t}$$

But $\frac{\Delta x}{\Delta t} = v$, therefore:

$$\text{power} = Fv$$

> **STUDY TIP**
>
> A useful shortcut when calculating the efficiency of a system, if you know the power input and the power output, is to use the formula:
>
> efficiency =
>
> $$\frac{\text{useful power output}}{\text{power input}}$$

WORKED EXAMPLE

A cyclist is travelling along a straight, level road at a constant velocity of $27\,\text{km h}^{-1}$ against total frictional forces of $50\,\text{N}$. Calculate the power developed by the cyclist.

Answer

Convert the velocity from km h^{-1} into m s^{-1}:

$$27\,\text{km h}^{-1} = 27 \times \frac{1000}{3600} = 7.5\,\text{m s}^{-1}$$

$$\text{power} = \text{force} \times \text{velocity}$$

$$= 50 \times 7.5$$

$$= 375\,\text{W}$$

▶ NOW TEST YOURSELF

8 Calculate the power developed when a force of $3.0\,\text{N}$ moves its point of application through a distance of $6.4\,\text{m}$ in the direction of the force in a time of $30\,\text{s}$.

9 It takes a crane 3.0 minutes to lift a carriage of mass $2500\,\text{kg}$ through a height of $150\,\text{m}$. Calculate:
  a the work done against gravity
  b the average power output of the crane

10 A car is travelling at a steady $24\,\text{m s}^{-1}$ along a level road. The power output from the engine is $45\,\text{kW}$. Calculate the total frictional force on the car.

▶ REVISION ACTIVITIES

Make a flow chart to show how the units and/or dimensions of the quantities in the following list are linked.

acceleration   energy   work   force   length   mass   power   time

'Must learn' equations:

work done $= Fx\cos\theta$

$E_\text{p} = mg\Delta h$

$E_\text{k} = \frac{1}{2}mv^2$

efficiency of a machine $= \dfrac{\text{useful energy output}}{\text{total energy input}} \times 100\%$

$\text{power} = \dfrac{\text{work done}}{\text{time taken}} = \dfrac{\text{energy transformed}}{\text{time taken}}$

$\text{power} = Fv$

11 An incline is at an angle of 45° to the horizontal. A force of 25 N pulls a box 4.0 m along the incline. Calculate:

   a the total work done by the force

   b the work done against gravity by the force

   c the gravitational potential energy gained by the box

12 A ball of mass 0.30 kg initially at rest falls from a height of 25 m. It hits the ground at a speed of 18 m s^{-1}. Calculate:

   a the potential energy lost by the ball

   b the kinetic energy gained by the ball

   c the work done against friction

> ## END OF CHAPTER CHECK

In this chapter, you have learnt to: ☐

» understand the concept of work ☐

» recall and use the formula *work done = force × displacement* in the direction of the force ☐

» recall and understand that the efficiency of the system is the ratio of the energy output from a system (or the useful work done by the system) to the total input energy ☐

» use the concept of efficiency to solve problems ☐

» recall and understand the principle of conservation of energy ☐

» derive the formula $\Delta E_p = mg\Delta h$ for gravitational potential energy changes in a uniform gravitational field ☐

» recall and use the formula $\Delta E_p = mg\Delta h$ ☐

» derive the formula for kinetic energy $E_k = \frac{1}{2}mv^2$ ☐

» recall and use the formula $E_k = \frac{1}{2}mv^2$ ☐

» define power as work done (or energy transformed) per unit time ☐

» solve problems using $P = W/t$ ☐

» derive $P = Fv$ ☐

» recall and use $P = Fv$ to solve problems ☐

>> We have already seen how forces produce changes in the motion of objects; they can also change the shape of objects.
>> Forces in opposite directions will tend to stretch or compress an object.
>> If two forces tend to stretch an object, they are described as **tensile**.
>> If they tend to compress an object, they are known as **compressive**.

Elastic and plastic behaviour

KEY TERMS

Load is defined as the force that causes deformation of an object.

Extension is the difference in length of a spring when there is zero load and when a load is put on the spring.

Compression is the difference in length of a spring when a force is applied to the spring and when there is zero load.

Limit of proportionality is the point at which the extension (or compression) of a spring is no longer proportional to the applied load.

Forces on a spring

REVISED ☐

Figure 6.1(a) shows the apparatus used to investigate the extension of a spring under a tensile force. Figure 6.1(b) shows the results of an experiment.

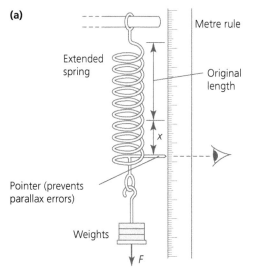

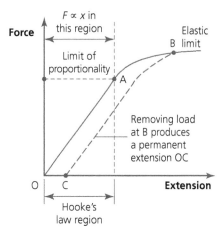

▲ Figure 6.1

Analysing the results, we see the following:

>> From O to A, the extension of the spring (x) is proportional to the applied force (F).
>> With larger forces, from A to B, the spring extends more easily and the extension is no longer proportional to the load.
>> When the force is reduced, with the spring having been stretched beyond point B, it no longer goes back to its original length.

From 0 to A, F is proportional to x:

$$F \propto x$$

This can be written as an equality by introducing a constant of proportionality:

$$F = kx$$

where k is the constant of proportionality, often known as the **spring constant**.

» The spring constant (k) is the force per unit extension.

$$k = \frac{F}{x}$$

» It is a measure of the stiffness of the spring.
» The larger the spring constant, the larger is the force required to stretch the spring through a given extension.
» The unit of the spring constant is **newton per metre** ($N\,m^{-1}$).

Point A, the point at which the spring ceases to show proportionality, is called the **limit of proportionality**. Very close to this point, there is a point B called the **elastic limit**. Up to the elastic limit, the deformation of the spring is said to be **elastic**.

If the spring is stretched beyond the elastic limit, it will not return to its original length when the load is removed. Its deformation is said to be **plastic**.

Hooke's law

REVISED ☐

Hooke's law sums up the behaviour of many materials that behave in a similar manner to a spring:

> Hooke's law states that the extension of an object is directly proportional to the applied force, provided the limit of proportionality is not exceeded.

Note that Hooke's law also applies to the compression of an object. In this case, the quantity x in the equation is the compression rather than the extension.

> **NOW TEST YOURSELF**   TESTED ☐

1 A spring has a length of 123 mm when there is no load. This increases to 135 mm when a load of 19.6 N is attached to it.
   The load is removed and a stone of unknown weight is attached to the spring, which now extends to a length of 145 mm.
   a Calculate the spring constant of the spring.
   b Calculate the weight of the stone. (Assume the elastic limit is not exceeded.)
2 A student investigates the extension of a spring. The results of the experiment are recorded in Table 6.1.

▼ Table 6.1

Mass of the load/g	Load/N	Length of the spring/cm	Extension/cm
0		26.2	0
100		28.3	
200		30.4	
300		32.8	
500		36.6	
600		38.8	

   a Copy and complete Table 6.1.
   b Plot a graph of load against extension.
   c Use the graph to determine the spring constant of the spring.

Check your answers at **www.hoddereducation.com/cambridgeextras**

Energy stored in a deformed material

REVISED ☐

Figure 6.2(a) shows the load–extension graph of an object that obeys Hooke's law. The work done in stretching the object is equal to the force multiplied by distance moved. This is equal to the elastic potential energy in the object. However, the force is not F, the maximum force – it is the average force, which is $\frac{1}{2}F$.

elastic potential energy, $E_p = \frac{1}{2}Fx$

This is the area of the triangle under the graph.

(a)
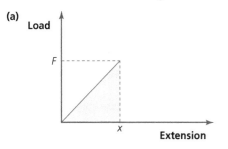

(b)
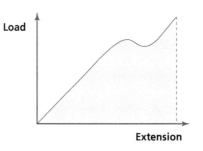

▲ Figure 6.2

This equation cannot be used with a material that has been extended beyond the limit of proportionality (Figure 6.2b) or any material that does not follow Hooke's law.

> **STUDY TIP**
>
> An alternative way to calculate the elastic potential energy stored in a deformed object is to use the spring constant rather than the force.
>
> $E_p = \frac{1}{2}Fx$
>
> But substituting for F from $F = kx$ gives:
>
> $E_p = \frac{1}{2}kx^2$

▶ **NOW TEST YOURSELF**   TESTED ☐

3   A load of 29.4 N, when attached to an unloaded spring, causes it to extend from 24.3 cm to 35.0 cm. Calculate the elastic potential energy stored in the stretched spring.
4   A spring has a length of 45 cm with no load on it. The spring is compressed so that its length is 18 cm. The spring has a spring constant of 24 N cm⁻¹. Calculate the elastic potential energy stored in the compressed spring.

Extension of a wire

REVISED ☐

» Applying tensile forces (or compressive forces) to a material will cause it to stretch (or compress).
» Metallic materials are often investigated by stretching a wire made from the material.
» For relatively small loads, the wire will obey Hooke's law.
» Up to the limit of proportionality, the wire will return to its original length when the load is removed.
» When the wire is stretched a little beyond the limit of proportionality, it will not return to its original length.
» The point at which the object no longer returns to its original shape is called the **elastic limit**.
» Extension up to the elastic limit is described as **elastic**.
» Extension beyond the elastic limit is described as **plastic**.

Stress and strain

Figure 6.3(a) shows the apparatus that could be used to investigate the stretching of a wire.

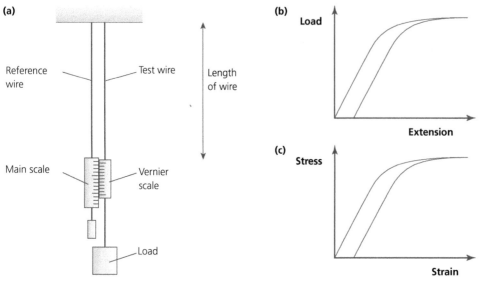

▲ Figure 6.3

The readings that need to be taken are shown in Table 6.2.

▼ Table 6.2

Reading	Reason	Instrument
Length of wire	Direct use	Metre rule
Diameter of wire	Enables the cross-sectional area to be found	Micrometer screw gauge
Initial and final readings from the vernier slide or a micrometer with a spirit level	The difference between the two readings gives the extension	Vernier scale Suitable device with a micrometer and a spirit level

» The graph obtained (Figure 6.3b) is similar to that obtained for a spring. This shows that Hooke's law may also be applied to stretching metals, up to their limit of proportionality.
» It is useful to draw the **stress–strain** graph (Figure 6.3c), which gives general information about a particular material, rather than for a particular Young modulus of the material.

Stress

REVISED

» From the definition, **stress** = normal force/area:

$$\sigma = \frac{F}{A}$$

» The formal symbol for stress is σ (the Greek letter sigma).
» The unit of stress is newton per metre squared or pascal (N m^{-2} = Pa).

KEY TERM

Stress is defined as the normal force per unit cross-sectional area of the wire.

Strain

REVISED

» From the definition, **strain** = extension/original length:

$$\varepsilon = \frac{x}{L}$$

» The formal symbol for strain is ε (the Greek letter epsilon).
» Strain is a ratio and is dimensionless so there are no units.

KEY TERM

Strain is the extension per unit of the unloaded length of the wire.

Check your answers at **www.hoddereducation.com/cambridgeextras**

Young modulus

The quantity stress/strain gives information about the elasticity of a material within the region of Hooke's law. This quantity is called the **Young modulus** (symbol E).

$$\text{Young modulus}, E = \frac{\text{stress}}{\text{strain}}$$

$$E = \frac{\text{force } (F)/ \text{ area } (A)}{\text{extension } (x)/ \text{ original length } (L)}$$

$$E = \frac{FL}{Ax}$$

The unit of the Young modulus is the same as for stress, i.e. the **pascal** (Pa).

WORKED EXAMPLE

A force of 250 N is applied to a steel wire of length 1.5 m and diameter 0.60 mm. Calculate the extension of the wire.
(Young modulus for mild steel = 2.1 × 10^{11} Pa)

Answer

cross-sectional area of the wire $= \pi\left(\frac{d}{2}\right)^2 = \pi\left(\frac{0.60 \times 10^{-3}}{2}\right)^2 = 2.83 \times 10^{-7} \text{m}^2$

Young modulus $= \dfrac{FL}{Ax}$

$2.1 \times 10^{11} = \dfrac{250 \times 1.5}{(2.83 \times 10^{-7}) \times x}$

$x = \dfrac{250 \times 1.5}{(2.83 \times 10^{-7}) \times (2.1 \times 10^{11})}$

$= 6.3 \times 10^{-3} \text{m} = 6.3 \text{mm}$

NOW TEST YOURSELF

TESTED

5 A wire of diameter 4.0 × 10^{-5} m supports a load of 4.8 N. Calculate the stress in the wire.

6 Calculate the strain on a wire of unstretched length 2.624 m which is stretched so that its length increases to 2.631 m.

> ## REVISION ACTIVITY
>
> 'Must learn' equations:
>
> $F = kx$
>
> $E_p = \frac{1}{2}Fx = \frac{1}{2}kx^2$
>
> $\text{Young modulus} = \dfrac{\text{stress}}{\text{strain}}$
>
> $\text{Young modulus} = \dfrac{FL}{Ax}$

END OF CHAPTER CHECK

In this chapter, you have learnt to:
» understand that deformation is caused by tensile and compressive forces
» understand and use the terms load, extension, compression and limit of proportionality
» recall and use the formula $F = kx$
» understand and use the term spring constant (k)
» determine the elastic potential energy of a material deformed within its limit of

proportionality from the area under a force–extension graph
» recall and use $E_p = \frac{1}{2}Fx = \frac{1}{2}kx^2$
» understand and use the terms elastic deformation, plastic deformation and elastic limit
» recall and use Hooke's law
» define and use the terms stress, strain and the Young modulus
» describe an experiment to determine the Young modulus of a metal in the form of a wire

7 Waves

In this course, you will meet various types of waves. The importance of waves is that they are a way of storing energy (**stationary waves**) and transferring energy from one place to another (**progressive waves**). Progressive waves transfer energy – they do not transfer matter.

Progressive waves

Terminology

A wave motion is formed when particles vibrate about their equilibrium position. There are many examples of wave motions, and a few examples you should be familiar with are: waves formed on springs, ropes, water (both in the natural world and on the surface of a ripple tank), sound waves and electromagnetic waves.

>> **Displacement** (x) of a particle is its distance from its equilibrium position. The unit is the metre (m).
>> **Amplitude** (x_0) is the maximum displacement of a particle from its equilibrium position. The unit is the metre (m).
>> **Phase difference** is the fraction of a cycle between two **oscillating** particles, expressed in either degrees or radians.
>> **Period** (T) is the time taken for one complete oscillation of a particle in the wave. The unit is the second (s).
>> **Frequency** (f) of a wave is the number of complete oscillations of a particle in the wave per unit time. The unit is the **hertz** (Hz).
>> **Wavelength** (λ) is the distance between points on successive oscillations of the wave that are vibrating exactly in phase. The unit is the metre (m).
>> **Wave speed** (v) is the distance travelled by the wave energy per unit time. The unit is the metre per second ($m\,s^{-1}$).

> **KEY TERMS**
>
> 1 **hertz** is one complete oscillation per second.
>
> An **oscillation** is one vibration of a particle – for example, from its mean position to the position of maximum displacement in one direction, back to the mean position, then to maximum displacement in the opposite direction and finally back to the mean position.

(a)

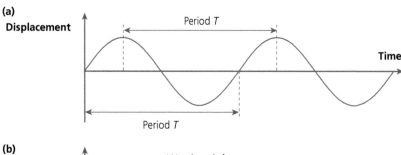

(b)

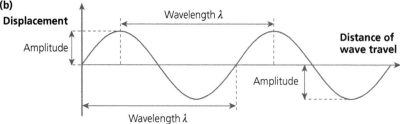

▲ **Figure 7.1**

Figure 7.1 shows (a) the displacement of a particle in a wave against time and (b) the displacement of all the particles at a particular moment in time.

> **STUDY TIP**
>
> It is easy to confuse these two graphs. Check the axes carefully. Figure 7.1(a) describes the variation of displacement with time, and Figure 7.1(b) describes the variation of displacement with position along the wave.

Phase difference

REVISED

Moving along a progressive wave, the vibrating particles are slightly out of step with each other, for example, there is a phase difference between points P and Q in Figure 7.2.

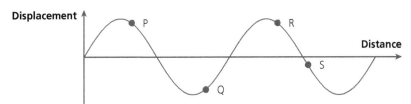

▲ **Figure 7.2 Phase difference**

Study Table 7.1, which describes the phase relationships between the different points on the wave in Figure 7.2.

▼ **Table 7.1**

Points	Phase difference/ degrees	Phase difference/ radians	Common terms used to describe the phase difference
P and R	360 or 0	2π or 0	in phase
P and Q	180	π	exactly out of phase (or antiphase)
R and S	90	$\frac{1}{2}\pi$	90° or $\frac{1}{2}\pi$ out of phase

Phase difference also compares how two sets of waves compare with each other. Figure 7.3 shows two sets of waves that are approximately 45° ($\frac{1}{4}\pi$) out of phase.

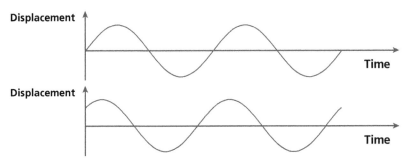

▲ **Figure 7.3**

> **STUDY TIP**
>
> In the AS Level course, you will use degrees to measure phase difference. You will meet radian measurement in 'Kinematics of uniform circular motion' on p. 115.

The wave equation

REVISED

The speed of a wave is given by the equation:

$$\text{speed } (v) = \frac{\text{distance}}{\text{time}}$$

By definition, in time T, the period of the oscillation, the wave travels one complete wavelength. Hence:

$$v = \frac{\lambda}{T}$$

But frequency and period are related by the equation:

$$f = \frac{1}{T}$$

Hence:

$$v = f\lambda$$

WORKED EXAMPLE

A car horn produces a note of frequency 280 Hz. Sound travels at a speed of 320 m s^{-1}. Calculate the wavelength of the sound.

Answer

$v = f\lambda$

$320 = 280 \times \lambda$

$\lambda = \dfrac{320}{280} = 1.14 \approx 1.1\,\text{m}$

> ## NOW TEST YOURSELF

TESTED ☐

1 State the relationship between the frequency of a wave and the time period of the oscillating particles in the wave.
2 Sketch a graph of the displacement against time of an oscillating particle in a wave. Describe how you could find the frequency of the wave.

Intensity of radiation in a wave

REVISED ☐

Progressive waves transfer energy. This can be seen with waves on the sea. Energy is picked up from the wind on one side of an ocean and is carried across the ocean and dispersed on the other side, as the wave crashes onto a shore.

Intensity is the energy transmitted per unit time per unit area at right angles to the wave velocity.

Energy transmitted per unit time is the power transmitted, so that:

$$\text{intensity} = \frac{\text{power}}{\text{area}}$$

The unit is watts per metre squared (W m^{-2}).

The intensity of a wave is proportional to the amplitude squared of the wave:

$$I \propto x_0{}^2$$

This means that if the amplitude is halved, the intensity is decreased by a factor of 4 (= 2^2).

> **KEY TERMS**
>
> **Intensity** is defined as the energy transmitted per unit time (power) per unit area at right angles to the wave velocity.

WORKED EXAMPLE

The intensity of light from a small lamp is inversely proportional to the square of the distance of the observer from the lamp, that is $I \propto 1/r^2$. Observer A is 1.0 m from the lamp; observer B is 4.0 m from the lamp. Calculate how the amplitude of the light waves received by the two observers compares.

Answer

intensity of light at B $= \left(\dfrac{1}{4}\right)^2 \left(= \dfrac{1}{16}\right)$ of that at A

intensity $\propto$ amplitude^2

amplitude $\propto \sqrt{\text{intensity}}$

amplitude at B $= \dfrac{1}{\sqrt{16}}$ of that at A

amplitude at B $= \dfrac{1}{4}$ of that at A

> **STUDY TIP**
>
> There are no simple formulae that you can apply here. You need to ensure that you understand the physics and then work through in a logical fashion.

Transverse and longitudinal waves

In mechanical waves, particles oscillate about fixed points. When a wave passes along a rope, the particles of the rope vibrate at right angles to the direction of transfer of energy of the wave. Water waves can also be considered to behave in a similar manner. This type of wave is called a **transverse wave** (see Figure 7.4).

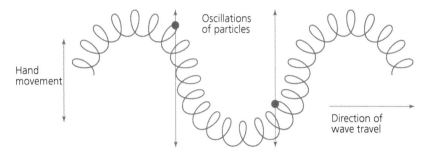

▲ Figure 7.4 Transverse wave

> **KEY TERMS**
>
> In a **transverse wave**, the particles vibrate at right angles to the direction of transfer of energy.

Sound waves are rather different.

» Particles vibrate back and forth parallel to the direction of transfer of energy of the wave.
» This causes areas where the particles are compressed together and areas where they are spaced further apart than normal.
» The areas where the particles are compressed together are called **compressions**.
» The areas where the particles are spaced further apart are called **rarefactions**.
» This type of wave is called a **longitudinal wave** (Figure 7.5).

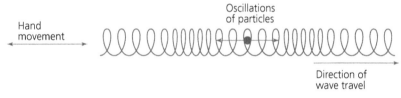

▲ Figure 7.5 Longitudinal wave

> **KEY TERMS**
>
> In a **longitudinal wave**, the particles vibrate parallel to the direction of transfer of energy.
>
> In a **compression**, the particles are closer together than normal.
>
> In a **rarefaction**, the particles are further apart than normal.

> ▶ **NOW TEST YOURSELF**   TESTED ☐
>
> 3 Figure 7.6 shows a graph of displacement against time of a particle in a longitudinal wave.
>
>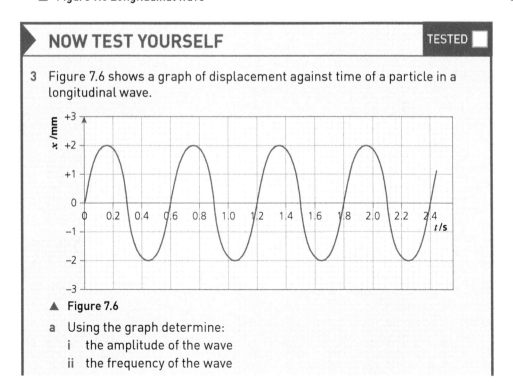
>
> ▲ Figure 7.6
>
> a Using the graph determine:
>   i the amplitude of the wave
>   ii the frequency of the wave

b Figure 7.7 shows the particles at rest before the wave arrives.

Direction of travel of the wave

▲ **Figure 7.7**

On a copy of Figure 7.7, show how one of the particles vibrates as the wave passes.

Sound waves

Measuring the frequency of a sound wave

- ➤ Sound waves are longitudinal waves.
- ➤ The frequency of a sound wave can be measured using a cathode-ray oscilloscope (CRO). The apparatus for this experiment is shown in Figure 7.8.

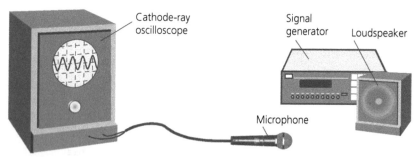

▲ **Figure 7.8 Measuring the frequency of a sound wave**

The period of the wave can be determined from the time-base setting and the number of oscillations shown on the screen (frequency = 1/period).

WORKED EXAMPLE

In Figure 7.8, the time-base is set at $5\,ms\,div^{-1}$. Calculate the frequency of the wave.

Answer

In four divisions, there are 3.5 oscillations.

Therefore, in $4 \times 5\,ms\,(= 20\,ms)$ there are 3.5 oscillations.

Therefore:

$$\text{period (time for one wave)} = \frac{20}{3.5}\,ms = 5.7 \times 10^{-3}\,s$$

$$f = \frac{1}{T} = \frac{1}{5.7 \times 10^{-3}} = 175\,Hz$$

STUDY TIP

When measuring the frequency of a wave with a CRO, use as much of the screen as possible to reduce uncertainties.

One period is from one peak (or one trough) to the next peak (or trough).

▶ NOW TEST YOURSELF

4 The engine of a train bumps into a row of trucks and a pulse moves down the row of trucks. Discuss whether it would be better to model the pulse as transverse in nature or longitudinal.

5 The signal from a signal generator is fed to a cathode-ray oscilloscope (CRO). The time-base is changed from $5\,ms\,div^{-1}$ to $50\,\mu s\,div^{-1}$. Discuss whether fewer or more oscillations are visible on the CRO screen.

Check your answers at **www.hoddereducation.com/cambridgeextras**

The Doppler effect

Listen to the pitch of a siren as a police car approaches and passes you. You will observe that on approach the pitch is higher than when the car is stationary and on leaving you the pitch is lower. This is known as the **Doppler effect.**

Figure 7.9 shows the wavefronts spreading from (a) a stationary source and (b) a moving source.

» Notice how the wavefronts from the moving source are much closer in front of the source, giving a shorter wavelength and higher frequency.
» Behind the source, the waves are further apart than normal, giving a longer wavelength and lower frequency.

(a)    **(b)**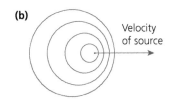
Velocity of source

▲ **Figure 7.9 (a) Waves spreading out from a stationary source, and (b) waves spreading out from a moving source**

The relationship between the observed frequency and the source frequency is given by the formula:

$$f_o = \frac{f_s v}{v \pm v_s}$$

where f_o is the observed frequency, f_s is the source frequency, v is the velocity of the waves and v_s is the relative velocity of the source and observer.

WORKED EXAMPLE

A loudspeaker connected to a signal generator produces a steady note of frequency 256 Hz. An observer moves towards the loudspeaker at a speed of $25\,\mathrm{m\,s^{-1}}$. Calculate the frequency of the sound that the observer hears (speed of sound = $330\,\mathrm{m\,s^{-1}}$).

Answer

$$f_o = \frac{f_s v}{v \pm v_s} = \frac{256 \times 330}{330 - 25} = 277\,\mathrm{Hz}$$

It is worth noting that it is the relative movement between an object and the source which is important, not the absolute movement.

> ## NOW TEST YOURSELF

TESTED ☐

6 An alarm is set off by a burglar. A nearby police officer runs as fast as he can towards the alarm. The burglar runs as fast as he can from the alarm. The householder remains in her garden.
   a Who hears the highest pitched sound?
   b Why would the change in pitch be very small?

KEY TERMS

The **Doppler effect** is the change in frequency of waves due to the relative motion of the wave source and the observer.

STUDY TIP

If the source and the observer are moving towards each other, the frequency increases and a minus sign is used in the denominator of the equation. If the source and observer are moving apart, a plus sign is used, leading to lower frequency.

STUDY TIP

All waves exhibit the Doppler effect. The electromagnetic radiation from galaxies shows a decrease in the frequencies in their spectra – known as the red shift. The fainter the galaxies, the greater the Doppler shift, which suggests that the further away the galaxy, the faster it is moving away from the Earth. This gives us evidence for the expansion of the Universe.

Electromagnetic waves

» Electromagnetic waves have the amazing property of being able to travel through free space (a vacuum).
» They are transmitted by the repeated variations in electric and magnetic fields.
» The electric and magnetic fields vibrate at right angles to the direction of travel of the disturbance.

» Electromagnetic waves are transverse in nature.
» You see light (a form of electromagnetic wave) that has travelled through billions of kilometres of empty space from distant stars.

Electromagnetic radiation comes at many different frequencies. Table 7.2 lists different types of electromagnetic radiation and their approximate wavelengths in a vacuum.

▼ Table 7.2 Types of electromagnetic radiation

Type of radiation	Approximate range of wavelength in a vacuum/m	Properties and uses
Gamma rays	10^{-16} to 10^{-11}	Produced by the disintegration of atomic nuclei; very penetrating, causes ionisation, affects living tissue
X-rays	10^{-13} to 10^{-9}	Produced from rapidly decelerated electrons; properties similar to gamma rays, the only real difference is in their method of production
Ultraviolet	10^{-9} to 4×10^{-7}	Ionising radiation, affects living tissue, stimulates the production of vitamin D in mammals
Visible light	4×10^{-7} to 7×10^{-7}	Stimulates light-sensitive cells on the retina of the eye
Infrared	7×10^{-7} to 10^{-3}	Has a heating effect and is used for heating homes and cooking
Microwaves	10^{-3} to 10^{-1}	Used in microwave cooking where it causes water molecules to resonate; also used in telecommunications, including mobile telephones
Radio waves	10^{-1} to 10^{5}	Used in telecommunications

» There are no sharp boundaries between these types of radiation. The properties gradually change as the wavelength changes. For example, it is not possible to give a precise wavelength at which radiation is no longer ultraviolet and becomes X-ray radiation.
» These radiations travel at the same speed, c, in a vacuum, where $c = 3.00 \times 10^8 \, \text{m s}^{-1}$. So, for calculations for electromagnetic waves, the wave equation can be written as $c = f\lambda$.
» If we know a radiation's frequency, we can calculate its wavelength in a vacuum.

WORKED EXAMPLE

The shortest wavelength that the average human eye can detect is approximately 4×10^{-7} m, which lies at the violet end of the spectrum. Calculate the frequency of this light.

Answer

$c = f\lambda$

$f = \dfrac{c}{\lambda}$

$f = \dfrac{3.0 \times 10^8}{4 \times 10^{-7}} = 7.5 \times 10^{14} \, \text{Hz}$

NOW TEST YOURSELF

TESTED ☐

7 A spectral line in the sodium spectrum has a frequency of 5.08×10^{14} Hz.
   a Calculate the wavelength of the light.
      (speed of light in a vacuum = $3.00 \times 10^8 \, \text{m s}^{-1}$)
   b Suggest the colour of this light.

Check your answers at **www.hoddereducation.com/cambridgeextras**

Polarisation

»» The oscillations in light waves are *not* generally in a single plane (as shown in Figure 7.10a,b), although some types of electromagnetic radiation do only have oscillations in a single plane. Figure 7.10(c) shows a vector diagram in which the unpolarised light is resolved into two perpendicular components.

»» When the oscillations are limited to a single plane (as with radio waves) the wave is said to be **plane polarised**.

(a)

Unpolarised light represented in three dimensions

(b)

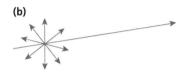

Vector diagram showing unpolarised light

(c)

Vector diagram showing the unpolarised light resolved into components at right angles

(d)

Vector diagram showing polarised light

> **STUDY TIP**
>
> When electromagnetic waves are polarised, the oscillating electric and magnetic fields are at right angles to each other. The vector diagram in Figure 7.10(d) is a simplification and represents the electric vector only.

▲ **Figure 7.10**

»» Some naturally occurring crystals polarise visible light, although a form of plastic polarisng filter known as Polaroid™ is more commonly used today.

»» Reflected light is partially polarised, thus polarising glasses are used to reduce glare.

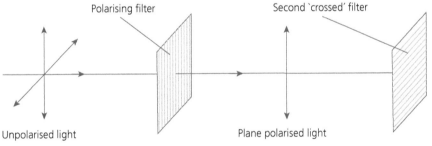

Polarising filter

Second 'crossed' filter

Unpolarised light

Plane polarised light

No light

▲ **Figure 7.11 The effect of crossed polarising filters**

»» Light can be polarised using a polarising filter.

»» Only oscillations in one direction pass through the filter.

»» A second polarising filter, with its axis at right angles to the first, will not allow any of the light through at all.

> **STUDY TIP**
>
> Longitudinal waves, where the vibrations are parallel to the direction of travel, cannot be polarised. Waves can be identified as transverse if they can be polarised.

Malus' law

REVISED ☐

»» Plane-polarised light of amplitude x_0 and intensity I_0 is incident on a polarising filter.

»» The light is polarised with the electric vector parallel to the filter's transmission axis, as shown in Figure 7.12(a). In this case, 100% of the radiation is transmitted.

»» The filter is then rotated, as shown in Figure 7.12(b). As it is rotated, the amount of radiation being transmitted reduces and the plane of polarisation rotates, always being parallel to the transmission axis.

» After a rotation of 90°, the radiation transmitted falls to zero.
» Figure 7.12(b) shows the filter after it has been rotated through an angle θ. The amplitude of the transmitted wave is now $x_0 \cos\theta$.

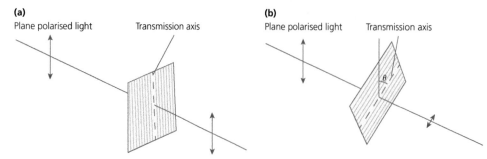

(a) Plane polarised light   Transmission axis

(b) Plane polarised light   Transmission axis

▲ **Figure 7.12**

» The intensity of a wave is proportional to the amplitude squared, therefore:

intensity of the transmitted wave $I = I_0 \cos^2\theta$

» This is known as Malus' law.
» Thus, we can calculate the intensity and hence the power transmitted when a plane polarised wave is incident on a polarising filter at any angle to the transmission axis of the filter.

WORKED EXAMPLE

The light from a laser pointer is plane polarised and has a power output of 5.0 mW. It passes through a polarising filter. The plane of polarisation is at an angle of 60° to the transmission axis of the filter. It then forms a spot on a wall of diameter 3.0 mm. Calculate the intensity of the spot of light on the wall.

Answer

Calculate the area of the spot:

$$\text{area} = \frac{\pi d^2}{4} = \pi \times \frac{(3.0 \times 10^{-3})^2}{4} = 7.1 \times 10^{-6}\,\text{m}^2$$

Calculate the intensity of the spot of light, ignoring the effect of the filter:

$$I = \frac{\text{power}}{\text{area}} = \frac{5 \times 10^{-3}}{7.1 \times 10^{-6}} = 710\,\text{W m}^{-2}$$

Use Malus' law to find the intensity of the light after passing through the filter:

$$I = I_0 \cos^2\theta = 710 \times \cos^2 60 = 180\,\text{W m}^{-2}$$

▶ REVISION ACTIVITIES

List the key terms in this chapter on pieces of card. Write out the meaning of each key term on separate pieces of card. Shuffle the two sets of cards and then try to match each key term with its definition.

'Must learn' equations:

$f = \dfrac{1}{T}$ intensity = power/area

$v = f\lambda$

$c = f\lambda$ $f_0 = \dfrac{f_s v}{(v \pm v_s)}$

$I = I_0 \cos^2\theta$

> ## NOW TEST YOURSELF
> TESTED ☐
>
> 8 A plane polarised beam of light is incident on a polarising filter with its plane of polarisation at an angle of 40° to its transmission axis. What fraction of the original intensity of the light is transmitted by the filter?

> ## END OF CHAPTER CHECK
>
> In this chapter, you have learnt:
> » that a wave motion is produced by particles vibrating about their mean positions ☐
> » the meanings of the terms displacement, amplitude, phase difference, period, frequency, wavelength and speed ☐
> » to derive the equation $v = f\lambda$ ☐
> » to recall and use the equation $v = f\lambda$ ☐
> » to understand that a progressive wave transfers energy without transferring matter ☐
> » to recall and use the equation: intensity = power/area ☐
> » to recall and use: intensity of a progressive wave $\propto$ (amplitude)^2 ☐
> » about the difference between transverse and longitudinal waves ☐
> » to analyse and interpret graphical representations of transverse and longitudinal waves ☐
> » to understand that when there is relative movement between a source of sound and
>
> an observer, there is a change in frequency of the sound, and this is known as the Doppler effect ☐
> » to use the expression $f_0 = f_s v/(v \pm v_s)$ ☐
> » to understand that electromagnetic waves are transverse ☐
> » to recognise that all electromagnetic waves travel at the same speed, c (3×10^8 m s^{-1}) in free space ☐
> » to recall and use the equation $c = f\lambda$ ☐
> » to recall the approximate range of wavelengths in free space of the different regions of the electromagnetic spectrum ☐
> » to recall that the wavelengths in the range 400–700 nm in free space are visible to the human eye ☐
> » that only transverse waves can be polarised ☐
> » to recall and use Malus' law ($I = I_0 \cos^2 \theta$) to calculate the intensity of plane polarised electromagnetic waves after transmission through polarising filters ☐

8 Superposition

Stationary waves

The principle of superposition

Superposition occurs when two waves of the same type meet. When the waves combine, we add the displacements, taking into account the direction of displacement.

The graphs (a) and (b) in Figure 8.1 show two waves at the moment in time when they meet. Readings of the displacements are taken from graphs (a) and (b) at 0.125 m intervals along the x-axis. The results are shown in Table 8.1.

KEY TERMS

The principle of **superposition** states that if two waves meet at a point, the resultant displacement at that point is equal to the algebraic sum of the displacements of the individual waves at that point.

▼ Table 8.1

Distance/m	Displacement of wave in graph (a)/cm	Displacement of wave in graph (b)/cm	Displacement of wave in graph (c)/cm
0	0	0	0
0.125	0.8	1.0	1.8
0.250	1.2	1.8	3.0
0.375	0.8	2.3	3.1
0.500	0	2.5	2.5
0.625	−0.8	2.3	1.5
0.750	−1.2	1.8	0.6
0.875	−0.8	1.0	0.2
1.000	0	0	0
1.125	0.8	−1.0	−0.2
1.250	1.2	−1.8	−0.6
1.375	0.8	−2.3	−1.5
1.500	0	−2.5	−2.5
1.625	−0.8	−2.3	−3.1
1.750	−1.2	−1.8	−3.0
1.875	−0.8	−1.0	−1.8
2.000	0	0	0

STUDY TIP

In practice, the uncertainties in this example are so great that much of the detail is lost. Both the frequency of samples taken and the precision with which the displacements are measured need to be increased.

These points can now be plotted onto the graph.

(a)

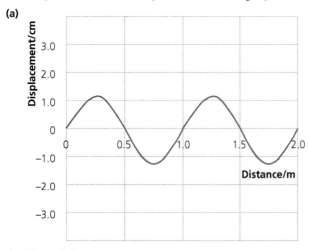

(b)

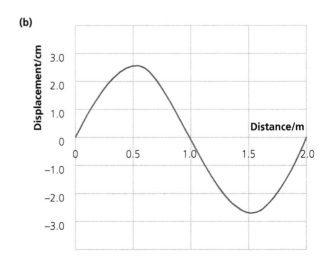

▲ Figure 8.1

Check your answers at **www.hoddereducation.com/cambridgeextras**

NOW TEST YOURSELF

TESTED ☐

1 Use the information in Table 8.1 to plot a graph of the resultant wave from combining waves (a) and (b)

Stationary waves in stretched strings

REVISED ☐

If you pluck a stretched string at its centre, it vibrates at a definite frequency, as shown in Figure 8.2.

>> This is an example of a stationary wave.
>> It is produced by the initial wave travelling along the string and reflecting at the ends.
>> This means that there are two waves of the same frequency travelling along the same line in opposite directions.
>> The two waves combine according to the principle of superposition.
>> This wave, where there is just a single loop, is called the **fundamental** wave or the **first harmonic**. Its wavelength is twice the length of the string.

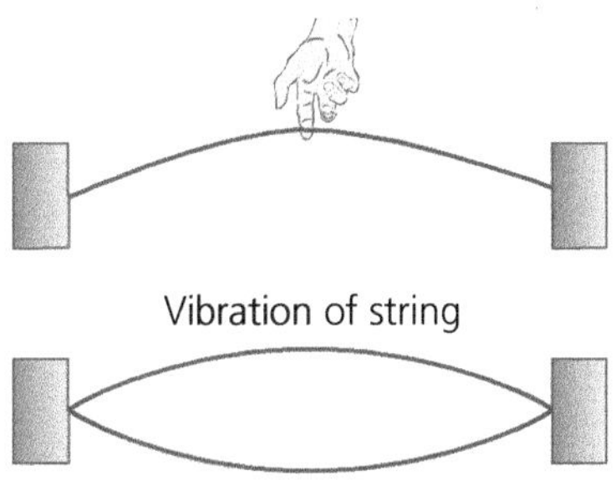

Vibration of string

▲ Figure 8.2 Fundamental wave, or first harmonic

Overtones and harmonics

>> A different stationary wave can be set up by plucking the string at points A and B (Figure 8.3).
>> The midpoint of the string has zero amplitude. This point is called a **node**.
>> The points of maximum amplitude are called **antinodes**.

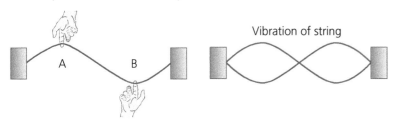

Vibration of string

▲ Figure 8.3 Second harmonic

> **KEY TERMS**
>
> A **node** is a point on a stationary wave that has zero displacement.
>
> An **antinode** is a point on a stationary wave that has maximum amplitude.

The frequency of this wave is twice that of the previous wave and its wavelength is half that of the fundamental. It is called the second harmonic or first overtone.

In general:

>> A wave travels down the string and is reflected back towards the other end; thus, two progressive waves of equal amplitude and frequency travel in opposite directions along the string.
>> The two waves will combine according to the principle of superposition.
>> Moving along the string, there are points where the two waves are in phase; at these points the oscillations add to form an antinode (maximum amplitude of oscillation occurs).
>> At those points where the waves are 180° out of phase, the oscillations subtract and a node is formed (minimum amplitude of oscillation occurs).
>> At the intermediate points along the string, the two waves combine producing oscillations between the maximum and minimum amplitudes.

Forced vibrations

Stationary waves die away quickly as energy is transferred to the surroundings. They can be kept going by feeding energy into the system (Figure 8.4). Stationary waves of this type are referred to as forced vibrations.

▲ Figure 8.4 A vibrator feeds energy into the system

For a stable stationary wave to be formed, the vibrator must have a frequency equal to the fundamental frequency (or a whole number times the fundamental frequency) of the string. Figure 8.5 shows the first three harmonics of a wave on a string.

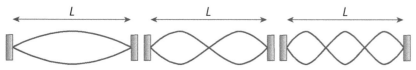

▲ Figure 8.5 A series of harmonics

STUDY TIP

Remember that the distance between adjacent nodes, or between adjacent antinodes, is half a wavelength, *not* a full wavelength.

Stationary waves in air columns

» Sound waves can produce stationary waves in air columns or pipes. The pipes may either be closed at one end and open at the other or open at both ends.
» A small loudspeaker connected to a variable frequency signal generator is used to feed energy into the system (Figure 8.6).
» As with vibrating strings, **resonance** occurs when the frequency of the sound produced by the loudspeaker is equal to a whole number times the fundamental frequency of the air column.

KEY TERM

Resonance is when the fundamental frequency of vibration of an object is equal to the driving frequency, giving a maximum amplitude of vibration.

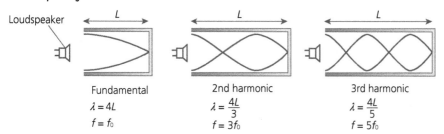

Fundamental	2nd harmonic	3rd harmonic
$\lambda = 4L$	$\lambda = \frac{4L}{3}$	$\lambda = \frac{4L}{5}$
$f = f_0$	$f = 3f_0$	$f = 5f_0$

▲ Figure 8.6 Stationary waves in a pipe with one end closed

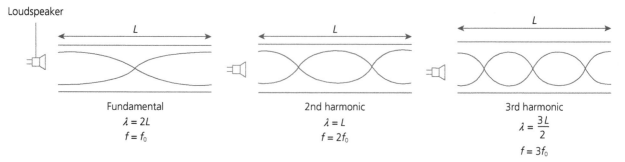

Fundamental	2nd harmonic	3rd harmonic
$\lambda = 2L$	$\lambda = L$	$\lambda = \frac{3L}{2}$
$f = f_0$	$f = 2f_0$	$f = 3f_0$

▲ Figure 8.7 Stationary waves in a pipe open at both ends

Note the difference in the wavelengths between the closed and open pipes in Figures 8.6 and 8.7. A fun exercise is to blow across the top of an open-ended tube and once you have got a note to sound, put your hand over the end (away from your lips) and listen for the change in pitch.

STUDY TIP

The diagrams with sound waves are graphical representations showing displacement against position along the tube. Students often interpret them as showing a transverse wave. Sound waves are longitudinal, so the displacement is parallel to the length of the tube *not*, as the diagrams can suggest, perpendicular to it. This is illustrated in Figure 8.8.

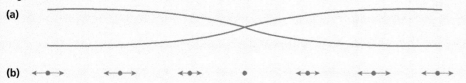

(a)

(b)

▲ Figure 8.8 (a) shows the graphical representation of a sound wave, and (b) represents the vibration of the particles

Differences between stationary waves and progressive waves

REVISED ☐

Some differences between stationary and progressive waves are given in Table 8.2.

▼ Table 8.2

Stationary waves	Progressive waves
Energy is stored in the vibrating particles	Energy is transferred from one place to another
All the points between successive nodes are in phase	All the points over one wavelength have different phases
The amplitudes of different points vary from a maximum to zero	All the points along the wave have the same amplitude

NOW TEST YOURSELF   TESTED ☐

2   The fundamental frequency of vibration of a guitar string is 128 Hz. When a source near to the guitar vibrates with a frequency of 640 Hz, the guitar string is observed to vibrate without being touched. Use your knowledge of stationary waves to explain this phenomenon.

Determining the speed of sound

REVISED ☐

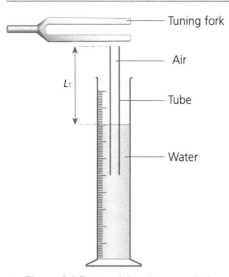

Tuning fork

Air

L_1

Tube

Water

▲ Figure 8.9 Determining the speed of sound

- » Apparatus for determining the speed of sound is shown in Figure 8.9.
- » The length of tube above the water is adjusted (starting from length = zero) until the fundamental stationary wave is formed. This can be identified by a clear increase in the loudness of the sound produced.
- » The length L_1 is measured.
- » The length of tube above the water is increased until the next stationary wave is formed. The new length L_2 is measured.
- » The wavelength is equal to $2(L_2 - L_1)$. If the frequency of the tuning fork is known, the speed of the sound in the air column can be calculated using the wave equation, $v = f\lambda$.

WORKED EXAMPLE

A tuning fork of frequency 297 Hz produces a stationary wave when a tube of air is 28.5 cm long. The length of the tube is gradually increased and the next stationary wave is formed when the tube is 84.0 cm long. Calculate the speed of sound in the tube.

Answer

$$\frac{\lambda}{2} = (84.0 - 28.5) = 55.5\,\text{cm}$$

$$\lambda = 111\,\text{cm} = 1.11\,\text{m}$$

$$v = f\lambda = 297 \times 1.11 = 330\,\text{m s}^{-1}$$

Determining the wavelength of microwaves

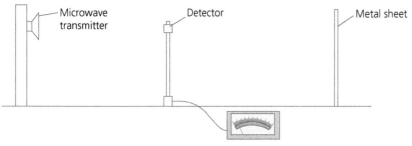

▲ Figure 8.10

- » A microwave transmitter is set up facing a metal sheet (Figure 8.10).
- » The metal sheet reflects the microwaves and produces stationary waves.
- » The distance between the transmitter and the metal sheet is adjusted so that as the detector is moved along the line between the transmitter and the metal sheet a series of maxima and minima are detected.
- » The distance between successive maxima (or successive minima) is equal to half the wavelength of the microwaves.

▶ NOW TEST YOURSELF

3 A child blows across the top of a pipe of length 40 cm. The pipe is open at both ends. A note of frequency 400 Hz is produced. Assume that the air column in the pipe is vibrating in its fundamental mode.
  - a Determine the wavelength of the stationary wave that is formed.
  - b Calculate the speed of sound in the air in the pipe.
4 A tuning fork of frequency 216 Hz is struck and held above the open end of a resonance tube.

The first resonance peak is produced when the tube length is adjusted to 361 mm. The second resonance peak is found when the tube has a length of 1065 mm. Calculate the speed of sound.
5 A microwave transmitter is positioned facing a metal plate. The plate is adjusted so that a stationary wave is formed between the plate and the transmitter. The distance between the first maximum and the twenty fifth maximum is 30 cm. Calculate the wavelength of the microwaves.

Check your answers at **www.hoddereducation.com/cambridgeextras**

Diffraction

>> When waves, for example, water waves in a ripple tank, pass through an aperture, they tend to spread out. This is known as **diffraction**.
>> Similarly, if waves travel past an object, they tend to spread around it.
>> Figure 8.11 shows **wavefronts** passing through a narrow aperture, through a wide aperture and around an object.
>> The degree of spreading when waves pass through an aperture depends on the wavelength of the wave and the width of the aperture. The longer the wavelength, the greater the spreading; the wider the aperture, the less the spreading.
>> For appreciable diffraction, the width of the aperture (or diameter of the object) must be of a similar order of magnitude to the wavelength of the wave.

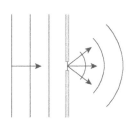 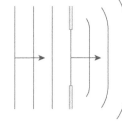 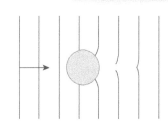

Small aperture   Large aperture

▲ **Figure 8.11 Wave diffraction in a ripple tank**

> ## NOW TEST YOURSELF   TESTED ☐

6 A harbour in a small fishing village is formed by building two harbour walls with a gap of about 20 m between them. Even when the boats are anchored within the shadow of the harbour walls, they are observed to bob up and down when the sea is rough. Explain why the boats bob up and down.

Interference

Interference   REVISED ☐

>> When two sets of waves of the same type meet, their displacements add or subtract, as explained by the principle of superposition.
>> At its most simple, when the two sets of waves are exactly in phase, the combined wave has an amplitude equal to the sum of the two amplitudes. This is known as **constructive interference** (see Figure 8.12).
>> When the two sets of waves are 180° out of phase (in **antiphase**) the two amplitudes subtract. This is known as **destructive interference** (see Figure 8.12). If the original amplitudes are equal, there will be no disturbance.
>> For interference to be observed, two **coherent** sources of waves are required.

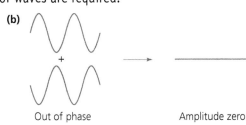

(a) In phase → Amplitude double
Constructive

(b) Out of phase → Amplitude zero
Destructive

▲ **Figure 8.12 Constructive and destructive interference**

Interference of sound

REVISED

Interference of sound waves can be demonstrated using two loudspeakers driven by the same signal generator, thus giving coherent waves (Figure 8.13).

▲ Figure 8.13 Inference of sound

» A loud sound is heard at A. The waves from the two loudspeakers have travelled equal distances and are in phase. The waves interfere constructively.
» At B a quiet sound is heard. The waves from the upper loudspeaker have travelled half a wavelength further than waves from the lower speaker. The waves are in antiphase, so they interfere destructively.
» At C a loud sound is heard. Waves from the upper loudspeaker have travelled a full wavelength further than waves from the lower speaker. The waves are now in phase and so interfere constructively.
» A quiet sound is heard at D as waves from the upper loudspeaker have travelled one-and-a-half wavelengths further than waves from the lower speaker. The waves are now in antiphase, so interfere destructively.

Interference of water waves

REVISED

Interference in water waves can be shown using a ripple tank (Figure 8.14). The areas of calm water (destructive interference) and rough water (constructive interference) can be viewed on the shadow image formed on the ceiling. Alternatively, they can be seen directly by looking almost parallel to the surface of the water.

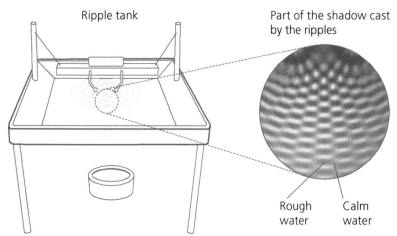

▲ Figure 8.14 Interference of water waves

Interference of light

REVISED

» Early attempts to demonstrate interference of light were doomed to failure because separate light sources were used.
» A lamp does not produce a single continuous wave train – it produces a series of short wave trains.
» The phase difference between one wave train and the next is random (Figure 8.15).

Check your answers at **www.hoddereducation.com/cambridgeextras**

» Hence, if light from two separate sources is mixed, there is no continuing relationship between the phases and an 'average brightness' is observed.

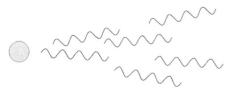

▲ Figure 8.15

To successfully demonstrate interference, the light from a single monochromatic source of light must be split and then recombined, with the two parts travelling slightly different distances (Figure 8.16).

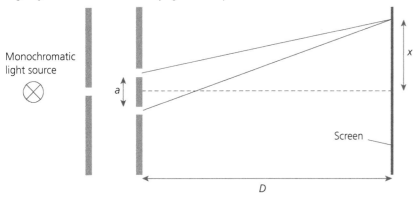

▲ Figure 8.16 Demonstrating interference with light from a single monochromatic source

The wavelength of light is very short ($\sim 10^{-7}$ m). Consequently, the distance between the slits a must be small (<1 mm) and the distance D from the slits to the screen must be large (≈ 1 m).

For constructive interference, the path difference between the contributions from the two slits is ax/D, where x is the distance between adjacent bright fringes.

Therefore:

$$\lambda = \frac{ax}{D}$$

WORKED EXAMPLE

Light of wavelength 590 nm is incident on a pair of narrow slits. An interference pattern is observed on a screen 1.5 m away. A student observes and measures 12 interference fringes, over a distance of 2.1 cm. Calculate the separation of the two slits.

Answer

$$\lambda = \frac{ax}{D}$$

$$x = \frac{2.1}{12}\,\text{cm} = 0.175\,\text{cm} = 1.75 \times 10^{-3}\,\text{m}$$

$$590 \times 10^{-9} = \frac{a \times (1.75 \times 10^{-3})}{1.5}$$

$$a = \frac{(590 \times 10^{-9}) \times 1.5}{1.75 \times 10^{-3}} = 5.1 \times 10^{-4}\,\text{m}$$

» The colour of light depends on its frequency.
» In general, for coherence, we use light of a single frequency, and therefore single wavelength. Light from a source producing a single frequency is called **monochromatic**.

> **KEY TERMS**
>
> **Monochromatic light** is light of a single frequency and, consequently, a single wavelength.

» In practice, when using white light (a whole range of colours), a few coloured fringes can be observed as the different wavelengths interfere constructively and destructively in different places.

Interference of microwaves

The principle is similar to the interference of light. The microwaves must be coherent and the output from a single source must be split into two sources using a pair of slits (Figure 8.17).

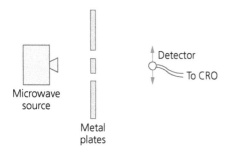

▲ **Figure 8.17 Interference of microwaves**

The slits are formed using metal plates, and the detector is positioned on the other side of the slits from the source.

> ### ▶ NOW TEST YOURSELF
> TESTED ☐
>
> 7 Copy Table 8.3. Use the formula $\lambda = ax/D$ to estimate the missing dimensions or wavelengths and complete the table.
>
> ▼ **Table 8.3**
>
> | Type of wave | Wavelength/m | D/m | x/m | a/m |
> |---|---|---|---|---|
> | Visible light | | 1 | 2×10^{-3} | 3×10^{-4} |
> | Sound waves | 1 | 5 | 4 | |
> | Microwaves | 3×10^{-2} | | 0.2 | 0.1 |
> | Ripples on a ripple tank | 1×10^{-2} | 0.5 | | 5×10^{-2} |
>
> 8 Laser light of wavelength 679 nm is incident on a pair of narrow slits. An interference pattern is observed on a screen 2.50 m away. The distance between the central and the 20th interference fringes is 12 cm. Calculate the separation of the slits.

Diffraction grating: multi-slit interference

The effect of using more than two slits to produce an interference pattern is to make the maxima sharper and brighter. The more slits there are, the sharper and brighter are the maxima. This makes it much easier to measure the distance between maxima.

The path difference between contributions from successive slits is $d\sin\theta$, where d is the distance between successive slits. Hence for a maximum:

$$n\lambda = d\sin\theta$$

where n is a whole number. The first maximum ($n = 1$) is sometimes called the first order (Figure 8.18). The central maximum is sometimes called the zeroth maximum.

> **STUDY TIP**
>
> The multi-slit device is called a diffraction grating, which is rather confusing. Although the spreading of the light (diffraction) is required for interference, this is really an interference grating.

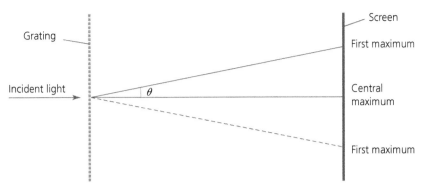

▲ **Figure 8.18**

WORKED EXAMPLE

Calculate the angles at which the first and second maxima are formed when a monochromatic light of wavelength 7.2×10^{-7} m is shone perpendicularly onto a grating with 5000 lines per cm.

Answer

$$d = \frac{1}{5000}\,\text{cm} = 2 \times 10^{-4}\,\text{cm} = 2 \times 10^{-6}\,\text{m}$$

For the first maximum:

$$\lambda = d \sin \theta$$

$$\sin \theta = \frac{\lambda}{d} = \frac{7.2 \times 10^{-7}}{2 \times 10^{-6}} = 0.36$$

$$\theta = 21°$$

For the second maximum:

$$2\lambda = d \sin \theta$$

$$\sin \theta = \frac{\lambda}{d} = \frac{2 \times (7.2 \times 10^{-7})}{2 \times 10^{-6}} = 0.72$$

$$\theta = 46°$$

▶ REVISION ACTIVITIES

'Must learn' equations:

$$\lambda = \frac{ax}{D}$$

$$n\lambda = d \sin \theta$$

Write down all the equations in this chapter and relate them to the situation in which they are used. For example, $n\lambda = d \sin \theta$ is used in multiple-slit diffraction.

▶ NOW TEST YOURSELF

TESTED ☐

9 A diffraction grating has 12500 lines per centimetre. When monochromatic light is shone on the grating, the first maximum is found to be at an angle of 30° to the central maximum. Calculate the wavelength of the light.

▶ END OF CHAPTER CHECK

In this chapter, you have learnt to:
- ▸ understand and use the principle of superposition ☐
- ▸ show an understanding of experiments that demonstrate stationary waves using microwaves, stretched strings and air columns ☐
- ▸ explain, using a graphical method, the formation of a stationary wave ☐
- ▸ identify nodes and antinodes ☐
- ▸ understand how wavelengths may be determined from the positions of nodes or antinodes of a stationary wave ☐
- ▸ explain the meaning of the term diffraction ☐
- ▸ understand experiments that demonstrate diffraction ☐

- ▸ understand the effect of the gap width relative to the wavelength as demonstrated by water waves in a ripple tank ☐
- ▸ understand the term interference ☐
- ▸ understand the term coherence ☐
- ▸ understand the experimental demonstration of interference using water waves, sound, light and microwaves ☐
- ▸ understand the required conditions for interference to be observed ☐
- ▸ recall and use the formula $\lambda = ax/D$ for double-slit inference of light ☐
- ▸ recall and use the formula $d \sin \theta = n\lambda$ ☐
- ▸ describe the use of a diffraction grating to determine the wavelength of light ☐

Electric current

Terminology

It is important to be clear about the meanings of the different terms used in electricity (Table 9.1).

▼ **Table 9.1 Terms used in electricity**

Quantity	Meaning	Unit and symbol
Current (I)	The flow of charge carriers	ampere (A)
Charge (Q)	'Bits' of electricity*	coulomb (C)
Potential difference (V)	The potential difference across a component is the energy transferred (or work done) as charge passes through the component	volt (V)
Resistance (R)	The opposition to current, defined as potential difference/current	ohm (Ω)

> **KEY TERMS**
> **Charge** passing a point = current × time for which the current flows.

*This is *not* a formal definition of charge. The concept of the nature of charge is quite complex. It can only be explained fully in terms of the interactions between charges, and between charges and electric fields. However, the reality is not that different. The smallest charge that is generally encountered is the charge on an electron (-1.6×10^{-19} C). We can consider charge to be 'bitty' in its nature. Physicists describe this as charge being quantised.

Current and charge carriers

» When a circuit is completed, the current is set up in the circuit almost immediately.
» The current front moves at (or near) the speed of electromagnetic radiation (3×10^8 m s^{-1}).
» It is wrong to think that the charge carriers (electrons in a metal) move at this speed.
» They move quite slowly, in the order of 0.1 mm s^{-1}, as they continually collide with ions in the crystal lattice. This is called their **drift velocity**.
» This can be compared with the high speed with which the wavefront of a longitudinal wave moves and the much smaller speeds at which the individual particles move.
» At any instant, each charge carrier will have a different velocity, and the drift velocity is the average of all these velocities.

Definitions of electrical units (1)

Current

The **ampere** is one of the base units described in the opening section of this book (p. 8). It is defined in terms of the number of electrons passing a point in a time of 1 second.

Charge

The charge that passes any point in a circuit when a current of 1 ampere flows for 1 second is 1 **coulomb**. This leads to equation 1:

$Q = It$ [equation 1]

Consider a conductor of cross-section A, through which there is a current I. In the conductor, there are n charge carriers per unit volume, each with a charge q and an average drift velocity v. In time t, the average distance travelled by each charge carrier = L (Figure 9.1).

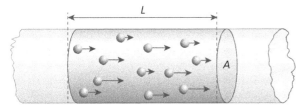

▲ **Figure 9.1**

Therefore:

$$L = vt$$

The volume of the shaded section of the conductor = AL = Avt.

All the charges (on average) initially in the shaded section of the conductor will pass through the cross-section labelled A in time t.

Therefore, the total charge passing through the cross-section A in time t = $Anvtq$.

charge passing per unit time $= \dfrac{Anvtq}{t} = Anvq$

charge passing per unit time = current I

So:

$$I = Anvq$$

WORKED EXAMPLE

A copper wire of cross-sectional area $4.0 \times 10^{-6}\,m^2$ carries a current of $2.5\,A$. The number of charge carriers per unit volume in copper is 8.4×10^{28}, each carrying a charge of $1.6 \times 10^{-19}\,C$. Calculate the mean drift velocity of the charge carriers.

Answer

$I = Anvq$

$v = \dfrac{I}{Anq}$

$= \dfrac{2.5}{(4.0 \times 10^{-6}) \times (8.4 \times 10^{28}) \times (1.6 \times 10^{-19})}$

$= 4.7 \times 10^{-5}\,m\,s^{-1}$

▶ **NOW TEST YOURSELF**   TESTED ☐

1 A battery is connected to a charger. It takes 2.5 hours to charge the battery using a steady current of $20\,mA$. Calculate the charge that flows through the battery.

2 Calculate the average current when a charge of $30\,C$ passes through a lamp in 3.0 minutes.

3 A carbon resistor of cross-sectional area $4.5 \times 10^{-5}\,m^2$ carries a current of $1.5\,A$. The mean drift velocity of the charge carriers is $2.3 \times 10^{-2}\,m\,s^{-1}$ and each charge carrier has a charge of $1.6 \times 10^{-19}\,C$. Calculate the number of charge carriers per unit volume in carbon.

Potential difference and power

Definitions of electrical units (2)

REVISED ☐

Potential difference

The unit of **potential difference** is the volt. There is a potential difference (p.d.) of 1 **volt** between two points when 1 joule of energy is transferred when a charge of 1 coulomb moves from one of the points to the other point. This leads to equation 2, where W is the energy transferred (or work done) in moving charge Q:

$$V = \frac{W}{Q} \quad \text{[equation 2]}$$

KEY TERMS

The **potential difference** between two points is the energy transferred per unit charge when charge is moved between the points.

Power

We met power on p. 48. You should remember that:

» power is defined as the work done, or energy transferred, per unit time
» the unit of power is the **watt** (W), which is the power generated when work is done at the rate of 1 joule per second

This leads to equation 3:

$$P = VI \quad \text{[equation 3]}$$

Resistance

A component has a resistance of 1 **ohm** (Ω) when there is a current of 1 ampere through the component and a potential difference of 1 volt across its ends.
This leads to equation 4:

$$R = \frac{V}{I} \quad \text{[equation 4]}$$

Summary of equations

$$Q = It \qquad V = \frac{W}{Q} \qquad P = VI \qquad R = \frac{V}{I}$$

The following relationships can be found by substituting the resistance equation into the power equation:

$$P = \frac{V^2}{R} \text{ and } P = I^2 R$$

WORKED EXAMPLE

A water heater of resistance $60\,\Omega$ runs from a mains supply of $230\,V$. It can raise the temperature of a tank of water from 20°C to 45°C in 20 minutes. Calculate:

a the charge that passes through the heater
b the energy dissipated by the heater

Answer

a $R = \dfrac{V}{I}$

$I = \dfrac{V}{R} = \dfrac{230}{60} = 3.83\,A$

$Q = It = 3.83 \times 20 \times 60 = 4600\,C$

b $V = \dfrac{W}{Q}$

$W = VQ = 230 \times 4600 = 1\,058\,000\,J \approx 1.1\,MJ$

STUDY TIP

The information about the temperature rise of the water is irrelevant to the question. One skill you need to develop is selecting relevant information and rejecting that which is not relevant.

> **NOW TEST YOURSELF**   TESTED ☐

4 A cell of 6.0 V and negligible internal resistance is connected across a
   resistor of resistance 4.0 Ω. Calculate:
   a the current through the resistor
   b the power dissipated in the resistor
   c the charge passing through the resistor in 15 minutes
   d the energy dissipated in the resistor in 15 minutes

Resistance and resistivity

I–V characteristics   REVISED ☐

The term resistance was defined on p. 78 as $R = \dfrac{V}{I}$, which leads to the equation $V = IR$.

Different components behave in different ways when there is a potential difference
across them. Examples are shown in Table 9.2.

▼ Table 9.2 *I–V* characteristics of various components

Component	Description		Explanation
Metal wire	The current is proportional to the potential difference across it. The resistance is the same for all currents, provided the wire remains at a constant temperature. The resistance of the wire is equal to the inverse of the gradient.	Metal wire	Metals contain many free electrons. These carry the current. The greater the potential difference, the greater the drift velocity of these electrons.
Filament lamp	At low currents, the current is proportional to the potential difference. At higher currents, the current does not increase as much for the same voltage increase. The resistance increases at higher currents.	Tungsten filament lamp	Lamp filaments are made from tungsten metal. At low currents, the filament behaves in the same way as the wire. At higher currents, the temperature increases to around 1500°C, the vibrations of the ions in the crystal lattice increase, presenting a larger collision cross-section and reducing the drift velocity of the electrons.
Thermistor	At low currents, the current is proportional to the potential difference. At higher currents, the current increases more for the same voltage increase. The resistance decreases at higher currents.	Thermistor	Like the metal wire, for low currents the potential difference is roughly proportional to the current, but when the temperature increases, the resistance decreases. Thermistors are semiconductors. Conduction in semiconductors is different from in metals. There are fewer free electrons. Increasing the temperature frees more electrons to carry the current and thus reduces the resistance.

→

Component	Description	Explanation
Diode	No current will pass in one direction. Once the potential difference (in the opposite direction) reaches a set value (0.6 V for a silicon diode), it conducts with very little resistance.	Diodes are also semiconductors but they are designed to allow currents to pass in one direction only.
Light-dependent resistor (LDR)	The resistance of a light-dependent resistor decreases as the intensity of the light falling on it increases.	As with thermistors, LDRs are semiconductors and have far fewer mobile charge carriers than metals. The energy from the light falling on the semiconductors frees the charge carriers so that there are more charge carriers to carry the current.

STUDY TIP

For a filament lamp, a thermistor, LDR and a diode, the resistance of the component is *not* equal to the inverse of the gradient. It is equal to the potential difference divided by the current when that p.d. is across the component.

Ohm's law

REVISED ☐

The special case of conduction through a metal is summed up in **Ohm's law**:

The current through a metallic conductor is proportional to the potential difference across the conductor provided the temperature remains constant.

▶ NOW TEST YOURSELF TESTED ☐

5 a A lamp has a potential difference of 3.0 V across it and there is a current of 1.2 A in the filament. Calculate the resistance of the lamp filament.

b The potential difference across the filament is increased to 12.0 V and the current increases to 3.6 A. Calculate the resistance of the filament now.

c Explain why the resistance of the filament changes in this manner.

Resistivity

REVISED ☐

The resistance of a component describes how well (or badly) a particular component or metal wire conducts electricity. It is often useful to describe the behaviour of a material; to do this, we use the idea of **resistivity**.

The **resistance** of a wire is:

» directly proportional to its length, $R \propto l$
» inversely proportional to the cross-sectional area: $R \propto \dfrac{1}{A}$

So:

$$R \propto \frac{L}{A}$$

Check your answers at **www.hoddereducation.com/cambridgeextras**

Hence:

$$R = \frac{\rho L}{A}$$

where ρ is the constant of proportionality, which is called the resistivity.

The units of resistivity are $\Omega\,m$:

$$\rho = \frac{RA}{L} \rightarrow \frac{\Omega\,m^2}{m} = \Omega\,m$$

WORKED EXAMPLE

A student wants to make a heating coil that will have a power output of 48 W when there is a potential difference of 12 V across it. The student has a reel of nichrome wire of diameter 0.24 mm. The resistivity of nichrome is $1.3 \times 10^{-8}\,\Omega\,m$. Calculate the length of wire that the student requires.

Answer

The resistance of the coil can be found from the equation:

$$P = \frac{V^2}{R}$$

$$R = \frac{V^2}{P} = \frac{12^2}{48} = \frac{144}{48} = 3.0\,\Omega$$

Now:

$$R = \frac{\rho L}{A}$$

and:

$$A = \pi\left(\frac{d}{2}\right)^2$$

$$= \pi\left(\frac{24 \times 10^{-5}}{2}\right)^2 = 4.52 \times 10^{-8}\,m^2$$

$$R = \frac{\rho L}{A} \rightarrow L = \frac{RA}{\rho} = \frac{3.0 \times (4.52 \times 10^{-8})}{1.3 \times 10^{-8}} = 10.4\,m$$

PRACTICAL SKILL

The cross-sectional area of the wire cannot be measured directly – it is a derived quantity found by measuring the diameter and then calculating the area. The diameter of the wire must be measured with a micrometer – the uncertainty in a micrometer reading is 0.01 mm.

Measure the diameter at three different points along the wire and then take an average of the three readings. Remember to record *all* your readings when taking an average, even if they are the same value. For further information about averaging results, see 'Estimating uncertainties' on pp. 105–107.

► NOW TEST YOURSELF

TESTED ☐

6 A lamp is designed to take a current of 0.25 A when it is connected across a 240 V mains supply. The filament is made from tungsten of cross-sectional area $1.7 \times 10^{-9}\,m^2$. Calculate:

   a the resistance of the filament
   b the length of wire required to make the filament (resistivity of tungsten $= 9.6 \times 10^{-7}\,\Omega\,m$)

▶ REVISION ACTIVITIES

Make a flow chart to show how the units and/or dimensions of the following quantities are linked:

charge   current   length   potential difference   power   resistance
resistivity   time

'Must learn' equations:

$Q = It$ $\qquad R = \dfrac{V}{I}$

$P = VI$ $\qquad R = \dfrac{\rho L}{A}$

$V = \dfrac{W}{Q}$

▶ END OF CHAPTER CHECK

In this chapter, you have learnt to:
- » understand that an electric current is a flow of charge carriers ☐
- » understand that the charge on carriers is quantised ☐
- » recall and use $Q = It$ ☐
- » understand that n, the number density of charge carriers, is the number of charge carriers per unit volume ☐
- » use the expression $I = Anvq$ for a current-carrying conductor ☐
- » define potential difference across a component as the energy transferred (or work done) per unit charge ☐
- » recall and use the equation $V = W/Q$ ☐
- » recall and use the equations $P = VI$, $P = I^2R$ and $P = V^2/R$ ☐
- » define resistance ☐
- » recall and use the equation $V = IR$ ☐
- » sketch I–V characteristics of a metallic conductor at a constant temperature ☐
- » sketch I–V characteristics of a semiconductor diode ☐
- » sketch I–V characteristics of a filament lamp ☐
- » explain that the resistance of a metallic conductor increases as the current increases because its temperature increases ☐
- » understand that the resistance of a light-dependent resistor (LDR) decreases as the light intensity increases ☐
- » understand that the resistance of a thermistor decreases as the temperature increases ☐
- » state Ohm's law ☐
- » recall and use the equation $R = \rho L/A$ ☐

Check your answers at **www.hoddereducation.com/cambridgeextras**

10 D.C. circuits

Practical circuits

Signs and symbols

You should familiarise yourself with the circuit symbols shown in Figure 10.1.

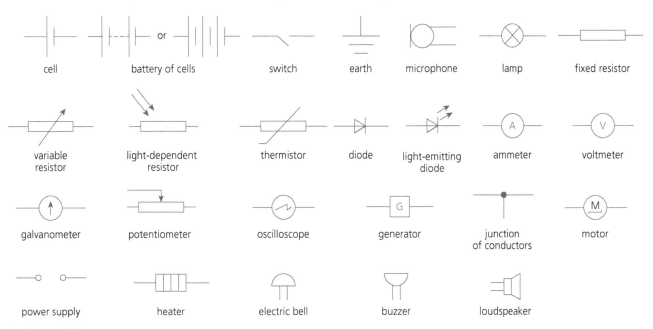

▲ Figure 10.1

Potential difference and e.m.f.

>> These two terms have similar but distinct meanings.
>> You have already met **potential difference** (p.d.) (p. 78). Remember that it is defined as the energy transferred, or work done, per unit charge when a charge moves between two points.
>> The term **e.m.f.** is used where a source of energy (such as a cell) gives energy to unit charge.
>> However, it is a little more precise than this. Feel a battery after it has delivered a current for some time – it is warm. This means that, as well as the battery giving energy to the charge, the charge is doing some work in overcoming resistance in the battery itself.
>> When e.m.f is defined, this work is included.

STUDY TIP

The term e.m.f. originally stood for electromotive force. This is rather confusing because it has nothing to do with force. Nowadays, e.m.f. stands as a term on its own.

KEY TERMS

The **potential difference** (p.d.) between two points is numerically equal to the energy transferred (or work done) per unit charge as a test charge moves from one point to the other.

The **e.m.f.** of a source is numerically equal to the energy transferred (or work done) per unit charge in driving charge around a complete circuit.

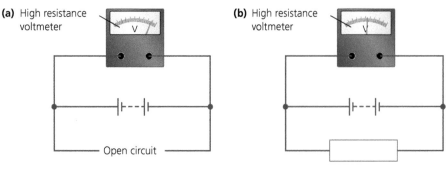

(a) High resistance voltmeter

(b) High resistance voltmeter

Open circuit

KEY TERM

The potential difference across a source of e.m.f., when there is a current through the source, is known as the **terminal potential difference**.

▲ **Figure 10.2 Potential difference across a battery**

➤➤ Figure 10.2(a) shows the potential difference when (virtually) no current is taken from the battery.

➤➤ This is (almost) equal to the e.m.f.

➤➤ Figure 10.2(b) shows how the potential difference across the battery falls when a current is taken from it. Some work is done driving the current through the battery.

Internal resistance

REVISED ☐

➤➤ You have seen how a source of e.m.f. has to do some work in driving a current through the source itself.

➤➤ In the case of a battery or cell, this is due to the resistance of the electrolytic solutions in the cell.

➤➤ In the case of a generator or transformer, it is due to the resistance of the coils and other wiring in the apparatus.

➤➤ It is clear that the source itself has a resistance; this is called the **internal resistance** of the source.

KEY TERM

The **Internal resistance** of a source of e.m.f. is the resistance inherent in the source itself as energy is transferred and charge is driven through the source.

It is often easiest to think of the two parts of a source of e.m.f. (the energy giver and the internal resistance) quite separately (Figure 10.3).

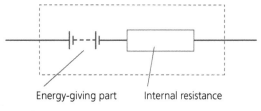

Energy-giving part   Internal resistance

▲ **Figure 10.3**

Consider a battery of e.m.f., *E*, and internal resistance, *r*, driving a current through an external resistance, *R*. The potential difference across the terminals of the battery is *V* (Figure 10.4).

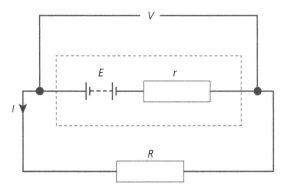

▲ **Figure 10.4**

When you work with internal resistances, treat them exactly the same as resistances in any other circuit. Work through the following equations to ensure that you understand the relationships.

$$E = I(R + r) = IR + Ir$$

But $IR = V$ and therefore:

$$E = V + Ir$$

WORKED EXAMPLE

A battery is connected across a resistor of $6.0\,\Omega$ and an ammeter of negligible resistance. The ammeter registers a current of $1.5\,A$. The $6.0\,\Omega$ resistor is replaced by an $18\,\Omega$ resistor, and the current falls to $0.6\,A$. Calculate the e.m.f. and internal resistance of the battery.

Answer

Consider the $6.0\,\Omega$ resistor:

$$E = IR + Ir = (1.5 \times 6.0) + 1.5r \rightarrow E = 9.0 + 1.5r$$

Consider the $18\,\Omega$ resistor:

$$E = IR + Ir = (0.6 \times 18) + 0.6r \rightarrow E = 10.8 + 0.6r$$

Substitute for E in the second equation:

$$9.0 + 1.5r = 10.8 + 0.6r$$

Therefore:

$$r = 2.0\,\Omega$$

Substitute for r in the first equation:

$$E = 9.0 + (1.5 \times 2) = 12\,V$$

> ## NOW TEST YOURSELF
> TESTED ☐
>
> 1 A battery has an e.m.f. of $1.4\,V$ and an internal resistance of $0.24\,\Omega$. A resistor of $3.5\,\Omega$ is connected across the terminals of the battery. Calculate:
>   a the current in the resistor
>   b the terminal potential difference across the battery

Kirchhoff's laws

Kirchhoff's first law
REVISED ☐

Kirchoff's first law is usually stated as follows:

> The sum of the currents entering any point in a circuit is equal to the sum of the currents leaving that point.

This is a restatement of the law of conservation of charge. It means that the total charge going into a point is equal to the total charge leaving that point.

WORKED EXAMPLE

Calculate the current I in Figure 10.5.

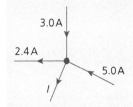

▲ Figure 10.5

Answer

Consider the currents going into the point as positive and those leaving the point as negative.

$$3.0 - 2.4 - I + 5.0 = 0$$

Therefore:

$$I = 5.6\,A$$

Kirchhoff's second law

Kirchoff's second law may be stated as:

> In any closed loop in an electric circuit, the algebraic sum of the electromotive forces is equal to the algebraic sum of the potential differences.

This is a restatement of the law of conservation of energy. Remember that potential difference between two points is the work done in moving unit charge from one point to the other. If the start point and the end point are the same, then the net energy change, or work done, must be zero.

Going around a loop, we consider instances where energy is given to the charge to be positive and where energy is lost by the charge to be negative.

WORKED EXAMPLE

Figure 10.6 shows a circuit. Calculate the e.m.f. of cell E_2 for the current through the ammeter to be zero.

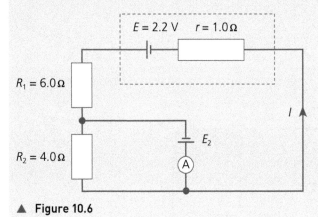

▲ **Figure 10.6**

Answer

Consider the outer loop and move anticlockwise around the loop:

$$2.2 - 6.0I - 4.0I - 1.0I = 0$$

$$I = 0.2\,A$$

Consider the inner loop, which contains the $4.0\,\Omega$ resistor and the cell, E_2. Again, move anticlockwise around the loop.

$$(-4.0 \times 0.2) - E_2 = 0$$

$$E_2 = -0.8\,V$$

The minus sign shows that in order to satisfy the conditions, the cell should be connected the other way around.

STUDY TIP

The e.m.f. of the second cell is put as $-E_2$ because the movement is from the positive to the negative cell – from a position of high potential energy to one of lower potential energy.

> **NOW TEST YOURSELF**
> TESTED

2 Figure 10.7 shows the currents at a junction in a circuit. Deduce the value of the current labelled I and give its direction.

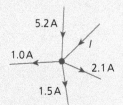

▲ **Figure 10.7**

3 A battery of e.m.f. 6.0V and an internal resistance of $1.6\,\Omega$ is connected across a resistor of resistance $4.8\,\Omega$. Use Kirchhoff's second law to show that the current through the resistor is 0.94A.

> **REVISION ACTIVITY**

Imagine how you would explain to a pre-AS Level physics student why (a) Kirchhoff's first law is a restatement of the law of conservation of charge and (b) Kirchhoff's second law is a restatement of the law of conservation of energy. Remember, they will almost certainly not have met these ideas before and may be a little hazy on their understanding of the terms potential difference, current and charge.

Check your answers at **www.hoddereducation.com/cambridgeextras**

Resistors in series

REVISED ☐

To find the total resistance of resistors connected in series (Figure 10.8), we can use Kirchhoff's second law.

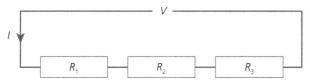

▲ **Figure 10.8 Resistors in series**

Going around the circuit:

$$V - IR_1 - IR_2 - IR_3 = 0$$

$$V = IR_1 + IR_2 + IR_3 = I(R_1 + R_2 + R_3)$$

But:

$$\frac{V}{I} = R_{total}$$

So:

$$R_{total} = R_1 + R_2 + R_3$$

Resistors in parallel

REVISED ☐

To find the total resistance of resistors connected in parallel (Figure 10.9), we can use Kirchhoff's laws.

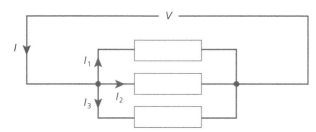

▲ **Figure 10.9 Resistors in parallel**

Using Kirchhoff's second law, we can see there is the same potential difference across each of the resistors, therefore:

$$I = \frac{V}{R_{total}} \qquad I_1 = \frac{V}{R_1} \qquad I_2 = \frac{V}{R_2} \qquad I_3 = \frac{V}{R_3}$$

Using Kirchhoff's first law:

$$I - I_1 - I_2 - I_3 = 0 \rightarrow I = I_1 + I_2 + I_3$$

Therefore:

$$\frac{V}{R_{total}} = \frac{V}{R_1} + \frac{V}{R_2} + \frac{V}{R_3}$$

And cancelling gives:

$$\frac{1}{R_{total}} = \frac{1}{R_1} + \frac{1}{R_2} + \frac{1}{R_3}$$

WORKED EXAMPLE

Figure 10.10 shows a network of resistors made up of five identical resistors each of resistance R. Calculate the total resistance of the network.

▲ Figure 10.10

Answer

resistance of the top line $= 2R$

resistance of the pair of resistors in parallel $= \left(\dfrac{1}{R_1} + \dfrac{1}{R_2}\right)^{-1} = 0.5R$

resistance of the lower line $= R + 0.5R = 1.5R$

$$\frac{1}{R_{total}} = \frac{1}{2R} + \frac{1}{1.5R} = \frac{3+4}{6R} = \frac{7}{6R}$$

$$R_{total} = \frac{6R}{7} = 0.86\,R$$

> ## NOW TEST YOURSELF
> TESTED ☐
>
> 4 A student has four resistors each of resistance $3.0\,\Omega$. Using all four resistors, draw circuits that could be used to make networks with resistance of:
>
> $12\,\Omega$ $0.75\,\Omega$ $4.0\,\Omega$ $3.0\,\Omega$ $1.2\,\Omega$

Potential dividers

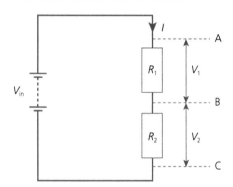

▲ Figure 10.11

A potential divider does exactly what the name suggests. Study Figure 10.11. If there is a potential V across AC then the total potential drop is divided between AB and BC.

In Figure 10.11, $V_1 = IR_1$ and $V_2 = IR_2$.

So:

$$\frac{V_1}{V_2} = \frac{IR_1}{IR_2} = \frac{R_1}{R_2}$$

A useful alternative way of working with this is:

$$V_{out} = \frac{R_2}{R_1 + R_2} V_{in}$$

where V_{out} is the potential drop across R_2 and V_{in} is the potential difference across the two resistors.

Check your answers at **www.hoddereducation.com/cambridgeextras**

WORKED EXAMPLE

Calculate the output potential difference in the circuit shown in Figure 10.12.

Answer

$$V_{out} = \frac{8.0}{16 + 8.0} \times 12 = 4.0\,V$$

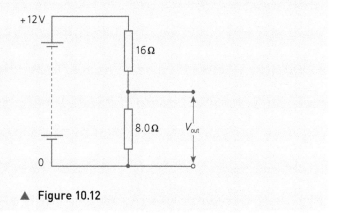

▲ Figure 10.12

Using a potential divider to provide a variable voltage output

REVISED

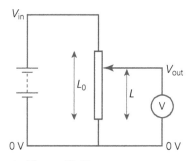

▲ Figure 10.13

The two resistors in a potential divider can be replaced by a single conductor, with a sliding contact to the conductor (Figure 10.13). The conductor could be a long straight wire, a strip of carbon or a coiled wire. Used in this way, the potential divider is called a **potentiometer**.

If a uniform wire is used, the output potential difference is:

$$V_{out} = \frac{L}{L_0} \times V_{in}$$

Using sensors in a potential divider

REVISED

Light-dependent resistor

» The resistance of an LDR decreases with increasing light levels (Figure 10.14a). Typical values range from $100\,\Omega$ in bright sunlight to in excess of $1\,M\Omega$ in darkness.
» A potentiometer with an LDR in one of the arms can be used to control the brightness of a lamp. In the circuit in Figure 10.15(a), the darker it gets, the greater the resistance of the LDR and the larger the potential difference across it. This can be used to automatically switch on lights.

Thermistor

» Although there are different types of thermistors, you only need to know about **negative temperature coefficient** thermistors.
» The resistance of a negative temperature coefficient thermistor decreases with increasing temperature (Figure 10.14b).
» The thermistor, connected into a potential divider, can be used to control the temperature. In the circuit in Figure 10.15(b), the output voltage increases as the temperature decreases. This can be used to increase the output from a heater.

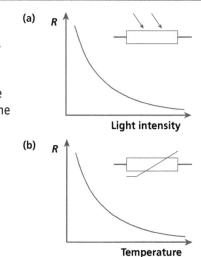

▲ Figure 10.14 Circuit symbol and characteristic of (a) an LDR and (b) a thermistor

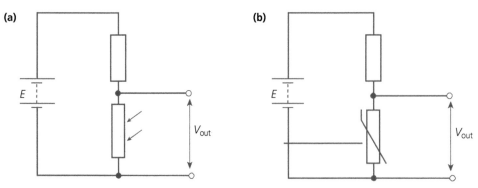

(a) E V_{out}

(b) E V_{out}

▲ **Figure 10.15 (a)** An LDR and **(b)** a thermistor connected into a potentiometer

Using a potential divider to compare potential differences

REVISED ☐

Comparing cells

When a potential divider is used to compare potential differences, it is usually referred to as a potentiometer.

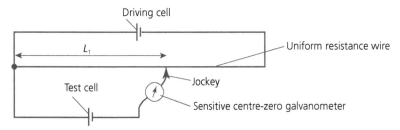

Driving cell

L_1

Uniform resistance wire

Test cell

Jockey

Sensitive centre-zero galvanometer

▲ **Figure 10.16**

» The circuit in Figure 10.16 can be used to compare the e.m.f. of two cells.
» The position of the jockey is adjusted until the current through the **galvanometer** is zero.
» The e.m.f. of the test cell (E_t) is now equal to the potential drop across the length L_1 of the resistance wire.
» This method of measurement is known as a **null method** – null meaning zero. The length L_1 is recorded.
» The test cell is then replaced with a standard cell of e.m.f. E_s. The position of the jockey is adjusted until the new null reading is found.
» The new length (L_2) is measured and recorded.
» The two e.m.f. values are related by the equation:

$$\frac{E_t}{E_s} = \frac{L_1}{L_2}$$

Comparing resistors

A similar method can be used to compare resistors.

» Two resistors are set up in series with a cell.
» The series circuit is then connected to the potentiometer, as shown in Figure 10.17, and the balance point is found ($L = L_1$).
» The potential drop across the resistor is IR_1.
» The leads from the potentiometer are disconnected and then reconnected across the second resistor (points B and C on the diagram).
» The new balance point is found ($L = L_2$).

> ### KEY TERMS
>
> A **galvanometer** is an instrument for detecting small currents or potential differences. A sensitive ammeter or voltmeter may be used as a galvanometer. They are often used in potentiometers, where a balance point and null reading are being looked for.
>
> A **null method** is one in which the apparatus is arranged so that a zero reading is required. The zero reading implies that the apparatus is balanced and that the value of an unknown can be found from the values of the constituent parts of the apparatus only.

Check your answers at **www.hoddereducation.com/cambridgeextras**

» The potential drop across this resistor = IR_2.
» Hence:

$$\frac{R_1}{R_2} = \frac{L_1}{L_2}$$

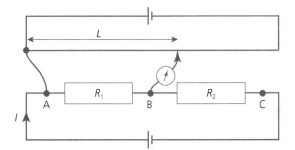

▲ Figure 10.17

WORKED EXAMPLE

A potentiometer, which has a conducting wire of length 1.0 m, is set up to find the e.m.f. of a dry cell. When the dry cell is connected to the potentiometer, the balance length is found to be 43.5 cm. A standard cell of e.m.f. 1.02 V is used to replace the dry cell. The balance length is now 12.9 cm less than for the dry cell. Find the e.m.f. of the dry cell.

Answer

The balance length for the standard cell = 43.5 − 12.9 = 30.6 cm.

$$\frac{E_t}{E_s} = \frac{L_1}{L_2}$$

$$\frac{E_t}{1.02} = \frac{43.5}{30.6} = 1.42$$

$$E_t = 1.42 \times 1.02 = 1.45 \, \text{V}$$

> **STUDY TIP**
>
> Work carefully through the development of the equations and ensure that you understand the logic of their development.

► NOW TEST YOURSELF

TESTED ☐

5 A potential divider is made up from a cell of e.m.f. 12.0 V and negligible internal resistance, and a wire of length 0.800 m and uniform thickness. Calculate the output voltage when the distance L_0 in Figure 10.13 on p. 89 is 0.430 m.

6 To find the e.m.f. of a cell, a scientist connects the cell to a potentiometer circuit similar to that in Figure 10.15. She adjusts the position of the jockey and gets a zero reading on the galvanometer when the length L_1 is 33.4 cm. She then disconnects the cell and replaces it with a standard cell of e.m.f. 2.71 V. A null reading is now found when L_2 is 74.3 cm. Calculate the e.m.f. of the original cell.

7 A resistor is connected in series with a standard resistor of resistance 12.0 Ω. They are connected to a potentiometer, as in Figure 10.16. A zero reading is found on the galvanometer when the distance L is 45.2 cm. The leads at A and B are disconnected and then reconnected across the standard resistor (points B and C). The null reading on the galvanometer is now found when L is 33.7 cm. Calculate the resistance of the unknown resistor.

> ## REVISION ACTIVITY

'Must learn' equations:

$$V_{out} = \frac{R_2}{R_1 + R_2} V_{in}$$

$$\frac{E_t}{E_s} = \frac{L_1}{L_2}$$

$$\frac{R_1}{R_2} = \frac{L_1}{L_2}$$

PRACTICAL SKILL

Measure the current in a component by placing the ammeter *in series* with the component.

Measure the potential difference *across* a component by placing a voltmeter *in parallel* with the component.

> ## END OF CHAPTER CHECK

In this chapter, you have learnt to:
» recall and use the circuit symbols used in this syllabus ☐
» draw and interpret circuit diagrams ☐
» define and use the e.m.f. of a source as energy transferred (or work done) per unit charge in driving charge around a complete circuit ☐
» distinguish between e.m.f. and potential difference (p.d.) in terms of energy considerations ☐
» understand the term terminal potential difference ☐
» understand the effects of the internal resistance of a source of e.m.f. on the terminal potential difference ☐
» recall Kirchhoff's first law ☐
» understand that Kirchhoff' first law is a result of the conservation of charge ☐
» recall Kirchhoff's second law ☐
» understand that Kirchhoff's second law is a result of the conservation of energy ☐

» derive, using Kirchhoff's laws, a formula for the combined resistance of two or more resistors in series ☐
» use the formula for the combined resistance of two or more resistors in series ☐
» derive, using Kirchhoff's laws, a formula for the combined resistance of two or more resistors in parallel ☐
» use the formula for the combined resistance of two or more resistors in parallel ☐
» understand the principle of a potential divider ☐
» explain the use of thermistors in potential dividers to provide a potential difference that is dependent on temperature ☐
» explain the use of light-dependent resistors in potential dividers to provide a potential difference that is dependent on light intensity ☐
» recall and use the principle of a potential divider as a means of comparing potential difference ☐
» understand the use of a galvanometer in null methods ☐

Atoms, nuclei and radiation

Nuclear model of the atom

Alpha-particle scattering experiment

In 1911, Rutherford's α-particle scattering experiment led to a model of the atom with a positively charged **nucleus** containing all the positive charge and virtually all the mass of the atom. This nucleus is surrounded by the much smaller, negatively charged **electrons**.

A plan view of the α-particle scattering apparatus is shown in Figure 11.1.

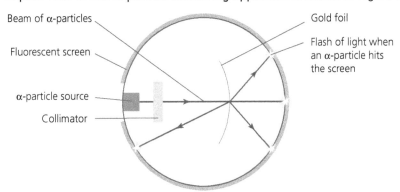

Beam of α-particles
Gold foil
Fluorescent screen
Flash of light when an α-particle hits the screen
α-particle source
Collimator

▲ Figure 11.1

Figure 11.2 shows the deflection of α-particles by the nucleus.

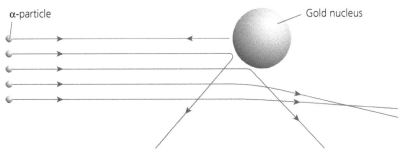

α-particle
Gold nucleus

▲ Figure 11.2

Results from the α-particle scattering experiment

» The vast majority of the α-particles passed straight through the gold foil with virtually no deflection.
» A few (approximately 1 in 10 000) of the α-particles were deflected through angles in excess of 90°.

Conclusions

» α-particles are positively charged with a mass about 8000 times that of an electron.
» The large-angle deflection could only occur if the α-particles interacted with objects more massive than themselves.
» This led Rutherford to develop the solar-system model of the atom (Figure 11.3).

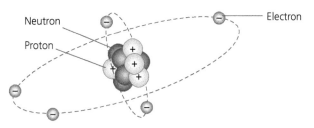

▲ Figure 11.3 Rutherford's solar-system model of the atom

STUDY TIP

Figure 11.3 is not drawn to scale. If it were, and the nucleus was kept to this size, the electrons would be over 100 m away.

» The small numbers of particles that are deflected through large angles indicate that the nucleus is very small.
» The proportions deflected in different directions enabled Rutherford to estimate the diameter of the nucleus as being in the order of 10^{-14} m to 10^{-15} m. This compares with an atomic diameter of about 10^{-10} m.

Subatomic particles

REVISED ☐

The nucleus of an atom has a structure made up of protons and neutrons. The subatomic particles are shown in Table 11.1.

▼ Table 11.1

Particle	Charge*	Mass**	Where found
Proton	$+1e$	$1u$	In the nucleus
Neutron	0	$1u$	In the nucleus
Electron	$-1e$	$1u/1840$	In the outer atom

*Charge is measured in terms of the electronic charge (e): $e = 1.6 \times 10^{-19}$ C.
**Mass is measured in unified atomic mass units (u): $1u$ is $\frac{1}{12}$ of the mass of a carbon-12 atom = 1.66×10^{-27} kg.

The chemical elements

REVISED ☐

» The different elements and their different chemical properties are determined by the number of protons in the nucleus.
» The number of protons, in turn, determines the number of electrons in the outer atom.
» The different atoms, or more precisely their nuclei, are fully described by the **proton number** and the **nucleon number**.

KEY TERMS

The **proton number** is the number of protons in a nucleus.

The **nucleon number** is the total number of protons plus neutrons in a nucleus.

Isotopes are different forms of the same element which have the same number of protons in the nuclei but different numbers of neutrons.

STUDY TIP

In some older books, you might see the proton number referred to as the atomic number and the nucleon number as the mass number. While these are not incorrect, it is better to use the up-to-date terms of nucleon number and proton number.

» All elements have **isotopes**. Isotopes have identical chemical properties but different physical properties.
» The chemical properties of different isotopes of the same element are identical because they have the same number of protons and hence the same electron configuration in the outer atom.
» A **nuclide** is fully described by the notation $_Z^A X$, as shown in Figure 11.4.

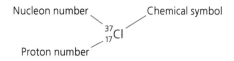

▲ Figure 11.4 Notation for chlorine-37

KEY TERMS

A **nuclide** is a single type of nucleus with a specific nucleon number and a specific proton number.

» This nuclide is an isotope of chlorine.
» The nucleus contains 17 protons and $(37 - 17) = 20$ neutrons.
» This isotope makes up about 24% of naturally occurring chlorine. The other 76% is made up of the isotope $^{35}_{17}Cl$, which contains 18 neutrons.

> ### NOW TEST YOURSELF
> TESTED ☐
>
> 1 Relative atomic masses are usually given as whole numbers, but chlorine is usually given as 35.5. Explain why the relative atomic mass of chlorine is given to this precision.
> 2 Consider the following isotopes.
>
> $$^{1}_{1}H \qquad ^{2}_{1}D \qquad ^{39}_{17}Cl \qquad ^{40}_{19}K \qquad ^{39}_{18}A$$
>
> a How many neutrons are there in each of the isotopes?
> b Which of the nuclides are isotopes of the same element?

Radioactive decay

REVISED ☐

Some nuclides are unstable and decay by emitting radiation; this is known as radioactive decay.

The four most common types of radioactive decay – alpha (α) decay, beta minus (β^-) decay, beta plus (β^+) decay and gamma (γ) decay – are shown in Table 11.2.

▼ Table 11.2

Type of particle	Nature	Charge	Penetration	Relative ionising power	Reason for decay
α	Fast moving helium nucleus (2 protons + 2 neutrons)	$+2e$	Weak penetration; is absorbed by a few centimetres of air, thin card or aluminium foil	High	Nucleus is too large; helium groupings form within the nucleus and sometimes escape
β^-	Very fast moving electron	$-e$	Fair; stopped by several millimetres of aluminium	Fair	Nucleus has too many neutrons; a neutron decays into a proton and an electron; the electron escapes from the nucleus
β^+	Very fast moving positron	$+e$	The free positron collides with an electron and the pair are annihilated producing a high-energy photon	Weak; see comment on penetration	Nucleus has too many protons; a proton decays into a neutron and a positive electron known as a positron; the positron escapes from the nucleus
γ	Short-wavelength electromagnetic radiation	Zero	High; only partly stopped by several centimetres of lead	Low	Usually emitted in conjunction with another event such as the emission of an α-particle as the nucleus drops back to a lower more stable energy state

» All particles have an antiparticle whose mass is the same as the particle but with opposite charge.
» The antiproton has a charge of $-1e$ and a mass of 1u.
» The antielectron – usually referred to as a **positron** – has the mass of an electron and a charge of $+1e$.

> ### NOW TEST YOURSELF
> TESTED ☐
>
> 3 Explain why the highly penetrative radiations cause little ionisation.

The different penetrating powers of the radiations can be explained by the relative ionising powers.

»» Each time radiation ionises a particle, it loses energy.
»» Thus, α-particles, which cause many ionisations per unit length, lose their energy in a much shorter distance than γ rays, which cause far fewer ionisations.

Nuclear reactions

You have already met radioactive decay in earlier courses and should be familiar with decay equations, such as that for americium decay when it emits an α-particle:

$$^{241}_{95}\text{Am} \rightarrow \,^{237}_{93}\text{Np} + \,^{4}_{2}\alpha + \text{energy}$$

Another common form of decay is β decay:

$$^{90}_{38}\text{Sr} \rightarrow \,^{90}_{39}\text{Y} + \,^{0}_{-1}\beta + \text{energy}$$

The third type of common decay is γ decay. This usually occurs following α emission as the remaining nucleons rearrange themselves into a lower energy state. For example, the decay of uranium-238:

$$^{238}_{92}\text{U} \rightarrow \,^{234}_{90}\text{Th} + \,^{4}_{2}\alpha + \,^{0}_{0}\gamma$$

Although α, β and γ decay are the most common forms of decay, there are many other possibilities. An important example is the formation of carbon-14 in the atmosphere. A neutron is absorbed by a nitrogen nucleus, which then decays by emitting a proton:

$$^{14}_{7}\text{N} + \,^{1}_{0}\text{n} \rightarrow \,^{14}_{6}\text{C} + \,^{1}_{1}\text{p}$$

> **STUDY TIP**
>
> You should recognise that both the nucleon number and the proton number are conserved in every decay.

WORKED EXAMPLE

A $^{16}_{8}\text{O}$ nucleus absorbs a neutron. The newly formed nucleus subsequently decays to form a $^{17}_{9}\text{F}$ nucleus.

a   Write an equation to show the change when the neutron is absorbed.

b   Deduce what type of particle is emitted when the decay of the newly formed nucleus occurs.

Answer

a $^{16}_{8}\text{O} + \,^{1}_{0}\text{n} \rightarrow \,^{17}_{8}\text{O}$

b   The equation for the decay may be written as:

$$^{17}_{8}\text{O} \rightarrow \,^{17}_{9}\text{F} + \,^{y}_{z}\text{X}$$

For the conservation of nucleon number (the superscripts), the value of y must be 0.

For the conservation of proton number (the subscripts), the value of z must be –1.

Thus, the particle emitted must be a β⁻ particle.

▶ NOW TEST YOURSELF

4   The isotope $^{142}_{58}\text{Ce}$ decays by α emission. Give an equation for this decay.

5   The isotope $^{90}_{39}\text{Y}$ decays by β⁻ emission. Give an equation for this decay.

Check your answers at **www.hoddereducation.com/cambridgeextras**

β-decay and the neutrino

>> Look at the energy distributions (in Figure 11.5) of α-particles when a sample of a specific isotope decays.
>> Compare it with that of the energy distributions when a sample of a different isotope decays by β decay.

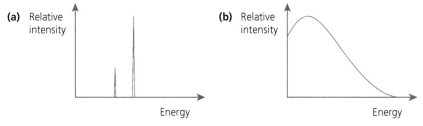

▲ **Figure 11.5 Energy spectra of (a) α emission and (b) β emission**

>> The α-particles have specific energies whereas the β-particles have a continuous range of energies.
>> The evidence from the α-spectrum is that the energies within the nucleus are quantised, in a similar way to the electrons in discrete energy levels in the outer atom.
>> So why is the β distribution continuous? This is explained by a second particle being emitted in β decay.
>> This particle has no charge and no (or very small) rest mass.
>> It is called the **neutrino**.
>> The β particle and the neutrino share the available energy in different proportions.

There is a balance in the matter and antimatter that is emitted, so that in β⁻ decay an antineutrino is emitted and in β⁺ decay a neutrino is emitted.

The equation for β⁻ decay is as follows:

$$^{1}_{0}\text{n} \rightarrow \, ^{1}_{1}\text{p} + \, ^{0}_{-1}\text{e} + \, ^{0}_{0}\bar{\nu} \qquad (\bar{\nu} \text{ is the symbol for the antineutrino})$$

The equation for β⁺ decay is as follows:

$$^{1}_{1}\text{p} \rightarrow \, ^{1}_{0}\text{n} + \, ^{0}_{1}\text{e} + \, ^{0}_{0}\nu \qquad (\nu \text{ is the symbol for the neutrino})$$

> **KEY TERMS**
>
> The **neutrino** is a particle that is emitted in β⁺ decay. It has zero charge and zero (or very little) rest mass.
>
> The **antineutrino** is the antiparticle of the neutrino and is emitted in β⁻ decay.

Fundamental particles

Leptons, hadrons and quarks

Leptons

>> Electrons, positrons, neutrinos and antineutrinos are believed to be fundamental particles with no further structure.
>> They are classed as **leptons** (meaning 'light ones').
>> Both electrons and neutrinos and their antiparticles are leptons.

Hadrons

>> Protons, neutrons and their antiparticles are known as **hadrons** ('heavy ones').
>> They are not fundamental and have an internal structure.
>> There are many different types of hadrons, with each type made up of different combinations of two or three smaller particles, which are called **quarks**.
>> Quarks are believed to be fundamental particles.
>> All hadrons have a charge that adds up to a whole number times the charge on an electron.

- There are six types (or 'flavours') of quark: up, down, charm, strange, top and bottom.
- Each flavour of quark has different properties such as upness, downness or strangeness.
- Each flavour has both a particle and an antiparticle.

Types of quark

REVISED ☐

In this course, we concentrate mainly on the quarks that make up protons and neutrons, the up and down quarks and their antiparticles.

▼ Table 11.3

Name	Symbol	Charge	Strangeness
up	u	$+\frac{2}{3}e$	0
down	d	$-\frac{1}{3}e$	0
antiup	ū	$-\frac{2}{3}e$	0
antidown	d̄	$+\frac{1}{3}e$	0
charm	c	$+\frac{2}{3}e$	0
strange	s	$-\frac{1}{3}e$	−1
anticharm	c̄	$-\frac{2}{3}e$	0
antistrange	s̄	$+\frac{1}{3}e$	+1
top	t	$+\frac{2}{3}e$	0
bottom	b	$-\frac{1}{3}e$	0
antitop	t̄	$-\frac{2}{3}e$	0
antibottom	b̄	$+\frac{1}{3}e$	0

Baryons and mesons

Proton and neutrons are **baryons**, which means they are each made up of three quarks. Figure 11.6 shows the quark structure of the proton and the neutron.

(a)   **(b)**

▲ **Figure 11.6 Quark structure of (a) a proton and (b) a neutron**

You can see that the charge on the proton is 1e; the charges on its three quarks are $+\frac{2}{3}e$, $+\frac{2}{3}e$ and $-\frac{1}{3}e$.

WORKED EXAMPLE

By considering the quarks that make up a neutron, show that its charge is zero.

Answer

A neutron has two down quarks, each with charge $-\frac{1}{3}e$ and an up quark, with charge $+\frac{2}{3}e$.

$$-\tfrac{1}{3}e - \tfrac{1}{3}e + \tfrac{2}{3}e = 0$$

Mesons are a second type of hadron and are made up of one quark and one antiquark. For example, the kaon (K⁺) consists of an up quark and an antistrange quark. This is written as us̄.

> ## NOW TEST YOURSELF
>
> **6** A baryon known as a charmed sigma consists of two up quarks and a charm quark. Determine the charge on the charmed sigma.
> **7** The quark structure of a pion is u$\bar{\text{d}}$
>   **a** Determine the charge on the pion.
>   **b** Suggest the quark structure of, and charge on, the antipion.

β decay and the quark model

$β^-$ decay occurs when a neutron decays into a proton. On the quark model, one of the down quarks converts into an up quark, releasing an electron and an antineutrino:

$$^1_0\text{n} \rightarrow \,^1_1\text{p} + \,^0_{-1}\text{e} + \,^0_0\bar{\text{v}}$$

» The total charge before and after the decay is the same.
» $β^-$ decay occurs when a neutron decays into a proton, an electron and an antineutrino.
» In terms of the quarks, a down quark turns into an up quark and an antineutrino.

$$-\tfrac{1}{3}\text{d} \rightarrow \tfrac{2}{3}\text{u} + -1\text{e} + 0\bar{\text{v}}$$

» Prior to the decay there is a down quark of charge $-\frac{1}{3}e$. After the decay, there is an up quark of $+\frac{2}{3}e$ and an electron of charge $-e$ and an antineutrino of zero charge, once more giving a total charge of $-\frac{1}{3}e$.
» The total charge is the same after the decay as before.

We have seen that in $β^-$ decay, a quark changes its flavour from up to down with the emission of an electron and an antineutrino.

> ## NOW TEST YOURSELF
>
> **8** In terms of quarks and other particles produced, suggest what changes occur in $β^+$ emission.
> **9** Suggest the quark composition of an antiproton.

> ## REVISION ACTIVITY
>
> Copy Tables 11.1, 11.2 and 11.3 onto thin card and keep them handy for reference. Laminate the cards if possible.

END OF CHAPTER CHECK

In this chapter, you have learnt:

» that the results from the α-particle scattering experiment show the existence of, and the small size of, the nucleus ☐

» to understand the simple model of the atom and to recognise that protons and neutrons make up the nucleus surrounded by orbital electrons ☐

» to understand the terms nucleon number and proton number ☐

» to understand that isotopes are forms of the same element with different numbers of neutrons in their nuclei ☐

» to understand and use the notation $^A_Z X$ for the representation of nuclides ☐

» to understand that nucleon number and proton number are conserved in nuclear processes ☐

» to describe the composition, mass and charge of α, β⁻, β⁺ and γ radiations ☐

» to understand that an antiparticle has the same mass but opposite charge to the corresponding particle ☐

» to recall that a positron is the antiparticle of an electron ☐

» to state that (electron) antineutrinos are produced during β⁻ decay and that neutrinos are produced during β⁺ decay ☐

» to understand that α-particles have discrete energies but that β⁻ particles have a continuous range of energies due to the emission of neutrinos/antineutrinos ☐

» to represent α and β decay by an equation of the form
$^{241}_{95}Am \rightarrow \, ^{237}_{93}Np + \, ^4_2\alpha$ ☐

» to use the unified atomic mass unit (u) as a unit of mass ☐

» to understand that a quark is a fundamental particle ☐

» to recall that there are six flavours of quark (up, down, charm, strange, top and bottom) ☐

» to recall and use the charge on each flavour of quark ☐

» to understand that the antiquark has the opposite charge to the respective quark ☐

» to recall that electrons and neutrinos are fundamental particles called leptons ☐

» to recall that protons and neutrons are not fundamental particles ☐

» to describe neutrons and protons in terms of their quark composition ☐

» to understand that hadrons may be either baryons or mesons ☐

» to recall that baryons consist of three quarks ☐

» to recall that mesons consist of a quark and an antiquark ☐

» to describe the changes to quark composition that take place during β⁻ and β⁺ decay ☐

Check your answers at **www.hoddereducation.com/cambridgeextras**

Experimental skills and investigations

The questions

Almost one-quarter of the marks for the AS Level examination are for experimental skills and investigations. These are assessed on Paper 3, which is a practical examination.

A total of 40 marks are available on Paper 3, divided equally between two questions. Although the questions are different on each Paper 3, the number of marks assigned to each skill is always similar.

The paper is designed to assess three skill areas:

»» manipulation, measurement and observation
»» presentation of data and observations
»» analysis, conclusions and evaluation

Manipulation, measurement and observation

Collection of data

REVISED ☐

You must be familiar with common laboratory apparatus, including rulers with millimetre scales, calipers, micrometer screw gauges, protractors, top-pan balances, newton meters, electrical meters (both analogue and digital), measuring cylinders, thermometers and stopwatches.

You will be expected to set up the apparatus supplied according to the instructions and diagram given on the question paper. The instructions detail the method to be followed and which measurements to take.

Variables

REVISED ☐

You should be familiar with the terms **independent variable** and **dependent variable** and be able to recognise them in different experiments. The following table shows some examples.

Investigation		Independent variable	Dependent variable
1	Investigating the height of a bouncing ball	Height from which the ball is dropped	Height to which the ball bounces
2	Investigating the period of vibration of masses suspended by a spring	Mass on the end of the spring	Periodic time
3	Investigating the melting of ice in water	Temperature of water	Time taken to melt
4	Investigating the current through resistors	Resistance of resistor	Current
5	Investigating e.m.f. using a potentiometer	e.m.f.	Balance length

We will refer to these examples later in the text, so you might like to put a marker on this page so that you can easily flip back as you read.

> **KEY TERMS**
> The **independent variable** is the variable that you control or change in an experiment.
>
> The **dependent variable** is the variable that changes as a result of the changing of the independent variable.

Range of readings

»» When you plan your experiment, you should use as wide a range of values for the independent variable as the equipment allows.
»» In general, the readings should be evenly distributed between the extremes of the range.
»» If you consider Investigation 3 in the table above – the melting ice experiment – the range of temperatures of the water in the beaker should be from nearly 100°C to about 10°C. You will probably be told how many readings to take, but it is likely to be a minimum of six sets. The values chosen for the independent variable should be taken at roughly equal intervals. A sensible spread might be 95°C, 80°C, 60°C, 45°C, 30°C and 15°C.
»» It sometimes makes sense to take several readings near a particular value – for example, if the peak value of a curved graph is being investigated. Practice in carrying out experiments will give you experience in deciding if this type of approach is necessary.

> **REVISION ACTIVITY**
>
> ....
>
> Accuracy, precision and uncertainty, including combining uncertainties, are discussed in detail in the 'Errors and uncertainties' section on pp. 10–13. You should refer back to this to refresh your memory.

Presentation of data and observations

Table of results

The table must be drawn before collecting results. It must have a sufficient number of columns to record the independent and dependent variables, as well as any calculated data. Each column must have a heading that includes the quantity and unit.

Raw data

The degree of precision of raw data in a column should be consistent. It will be determined by the measuring instrument used or the precision to which you can measure. This means that the number of significant figures may not be consistent. An example might be when measuring across different resistors using a potentiometer, where the balance points might vary from 9.3 cm to 54.5 cm.

Calculated data

With data calculated from raw measurements, the number of significant figures must be consistent with the raw measurements. This usually means that, except where they are produced by addition or subtraction, calculated quantities should be given to the same number of significant figures as (or one more than) the measured quantity of least precision. If a time is measured as 4.1 s, squaring this gives $16.81 \, s^2$. However, you would record the value as $16.8 \, s^2$ (or perhaps $17 \, s^2$). As with the raw data, this means that the number of significant figures in the column is not necessarily consistent.

The following table shows some readings from a resistance experiment and demonstrates how readings should be set out with:

»» the column headings, with quantity and unit
»» the raw data to the same precision
»» the calculated data to the relevant number of significant figures

p.d./V	Current$_1$/A	Current$_2$/A	Average current/A	Resistance/Ω
1.21	0.301	0.304	0.303	3.99
1.46	0.351	0.349	0.350*	4.17
1.73	0.358	0.364	0.361	4.79

* Do not forget to include the zero, to show that the current has been measured to the nearest milliamp.

Check your answers at **www.hoddereducation.com/cambridgeextras**

Graphs

Reasons for plotting graphs

Graphs:

» tend to average data, thereby reducing the effects of random errors
» identify anomalous points (which should then be investigated further)
» give information that can be used to identify relationships between variables

Rules for plotting graphs

1 **Draw and label axes**. Axes should be labelled with the quantity and the unit in a similar manner to column headings in a table. In general, the independent variable (the one you control) is put on the horizontal axis (*x*-axis). The dependent variable (the one that changes due to changes in the independent variable) goes on the vertical axis (*y*-axis).

2 **Choose sensible scales**. Scales should be chosen so that the points occupy at least half the sheet of graph paper used. However, awkward scales (1:3, 1:7, 1:11, 1:13 or their multiples) must be avoided. You do not necessarily have to include the origin on the graph if this means that a better use of the graph paper can be achieved.

3 **Plot points accurately**. Points should be plotted by drawing a small saltire cross (×) with a sharp pencil. Do *not* use dots or blobs. Blobs will be penalised and dots are often difficult to see as they tend to disappear into the line.

4 **Draw the best-fit straight line or best smooth curve**. The positioning of the points will show a trend, for example, pressure of a gas decreases as the volume increases. The trend line may be a curve or a straight line. When you draw a straight line, use a 30 cm ruler and a sharp pencil. There should be an equal number of points above and below the line. Take care that those points above and those below the line are evenly distributed along the line. Curves should be smooth and generally will form a single curve.

5 **Identify and check any anomalous points**. If a point is well off the line, go back and check it. In all probability, you will have made an error, either in plotting the point or in taking the reading.

> **REVISION ACTIVITY**
>
> Make a list of the five rules for plotting graphs on a piece of card and check them off each time you do an experiment.

Do *not* adjust your straight line or curve so that it goes through the origin. There may be good reasons why the dependent variable is not zero when the independent variable is zero. Consider Investigation 2 in the table on p. 101 – the experiment to investigate the period of vibration of a mass on the end of a spring. The measurements of the mass on the spring do *not* make allowance for the mass of the spring itself.

The figure below shows a typical straight-line graph.

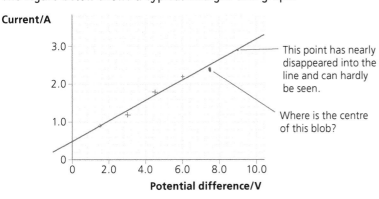

Curves should be drawn with a single sweep, with no feathering or sudden jerks. You need to practise doing this.

This figure shows two different curves drawn through the same set of points:

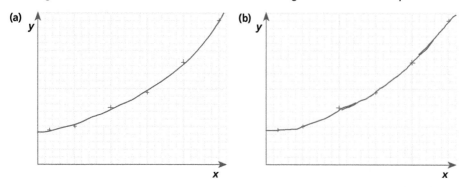

Graph (a) shows a well-drawn smooth curve. Graph (b) shows a poorly drawn curve through the same points. Note the jerkiness between the first two points and the feathering between points 2 and 4 and between points 5 and 6.

Analysis, conclusions and evaluation

Analysing a graph

REVISED ▢

Determining the gradient of a graph

The gradient of a graph is defined as:

$$\frac{\text{change in } y}{\text{change in } x} = \frac{\Delta y}{\Delta x} = \frac{y_2 - y_1}{x_2 - x_1}$$

When choosing the points to calculate the gradient, you should choose two points on the line (not from your table of results). To improve precision, the two points should be as far apart as possible and at least half the length of the drawn line.

In the straight-line graph on p. 103, two suitable points might be (0, 0.5) and (10.0, 3.2).

This gives a gradient of:

$$\frac{3.2 - 0.5}{10.0 - 0} = 0.27 \, \text{A V}^{-1}$$

The unit is not part of the gradient, but it will be essential when drawing conclusions about the quantity the gradient represents.

You may be asked to find the gradient of a curve at a particular point. In this case, you must draw a tangent to the curve at this point and then calculate the gradient of this line in a similar way to that described above.

Finding the y-intercept

The y-intercept of a graph is the point at which the line cuts the y-axis (that is, when $x = 0$). In the example on p. 103, the intercept is 0.5 A. When a false origin is used, it is a common mistake for students to assume that the vertical line drawn is the zero of x, so if you have used a false origin check carefully that the vertical line is at $x = 0$.

If, however, the chosen scale means that the y-intercept is not on the graph, it can be found by simple calculation.

»» Calculate the gradient of the graph.
»» The equation for a straight-line graph is $y = mx + c$. Choose one of the points used for calculating the gradient and substitute your readings into the equation.

WORKED EXAMPLE

The voltage input to an electrical device and the current through it were measured. A graph was drawn from the results.

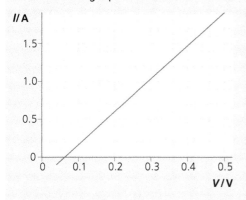

Determine the *y*-intercept on the graph.

Answer

Find the gradient. Use the points (0.06, 0) and (0.50, 1.90).

$$\text{gradient} = \frac{\Delta y}{\Delta x} = \frac{y_2 - y_1}{x_2 - x_1} = \frac{1.90 - 0}{0.50 - 0.06} = 4.3\,\text{A V}^{-1}$$

Substitute the first point and the gradient into the equation $y = mx + c$.

$$0 = (4.3 \times 0.06) + c$$

$$c = -0.26\,\text{A}$$

Conclusions

REVISED ☐

Determine the value of constants from a graph

Once the gradient and *y*-intercept have been determined, they can be used to find the values of constants in a given formula. In the previous worked example, the behaviour of the circuit can be expressed as:

$$I = SV + T$$

where I is the current, V is the potential difference, and S and T are two constants.

You must use the values of the gradient and the *y*-intercept to find the values of S and T and their units.

You will see that $I = SV + T$ is of the same form as $y = mx + c$.

Hence, constant S is the gradient (m) = 4.3 A V^{-1} and constant T is the intercept (c) = −0.26 A.

Evaluation

REVISED ☐

Estimating uncertainties

The precision of the instrument used for a measurement can be used to estimate the uncertainty; in instruments with analogue scales, this will usually be the smallest scale division. For example, when measuring the length of a wire with a metre rule that has millimetre divisions, the uncertainty will be ± 1 mm. A measured length of 25.0 cm would therefore be written as 25.0 ± 0.1 cm; the value of ± 0.1 cm is the absolute uncertainty. The percentage uncertainty is (0.1/25.0) × 100 = 0.4%.

For uncertainties in repeat readings, it is more useful to take half the range as the uncertainty. In the case of timing a pendulum, a student may take three measurements of time for ten swings of a particular length of the pendulum and average the reading. For example, the three readings may be 10.17 s, 10.04 s and 9.93 s.

The average = $\dfrac{10.17 + 10.04 + 9.93}{3}$ = 10.05 s.

The range is 10.17 − 9.93 = 0.24 so the absolute uncertainty is $\dfrac{0.24}{2}$ = 0.12 s.

The time can be written as 10.05 ± 0.12.

The percentage uncertainty is $\dfrac{0.12}{10.05}$ × 100 = 1.2%.

During an experiment, you should record any uncertainties in your measurements.

To ascertain if an experiment supports or fails to support a hypothesis, your result should lie within the limits of the percentage uncertainties. To support the hypothesis in the absence of any uncertainty calculations, a good rule of thumb is that the calculated value should lie within 10% of any predicted value.

The following worked examples take you through some of the stages of evaluating evidence.

WORKED EXAMPLE 1

In an initial investigation into the time it takes for an ice cube to melt in a beaker of water (Investigation 3 in the table on p. 101), the following results are obtained.

Trial 1: initial temperature of the water = 50°C

time taken to melt = 85 s

Trial 2: initial temperature of the water = 80°C

time taken to melt = 31 s

a Explain why it is only justifiable to measure the time taken for the ice cube to melt to the nearest second.

b Estimate the percentage uncertainty in this measurement in trial 1.

c Estimate the percentage uncertainty in this measurement in trial 2.

d Why is it more important to calculate the uncertainty in the time rather than in the initial temperature of the water?

Answer

a Even though the stopwatch that was used may have measured to the nearest one-hundredth of a second, it was difficult to judge when the last bit of ice disappeared.

b Suppose that the absolute uncertainty = ±5 s.
percentage uncertainty = $\pm \dfrac{5}{85}$ × 100% = 6%

c absolute uncertainty = ±5 s
percentage uncertainty = $\pm \dfrac{5}{31}$ × 100% = 16%

d The percentage uncertainty in measuring the temperature of the water is much less than the uncertainty in measuring the time. (±1°C, leading to ±1 to 2%).

» This example shows the reasoning in estimating the uncertainty in a measured quantity and how to calculate percentage uncertainty.

» You might feel that 5 s is rather a large uncertainty in measuring the time. It is at the upper limit, and you might be justified in claiming the uncertainty to be as little as 1 s.

» Nevertheless, if you try the experiment for yourself, and repeat it two or three times (as you should do with something this subjective), you will find that an uncertainty of 5 s is not unreasonable. The measurement of the initial temperature of the water has a much lower percentage uncertainty as less judgment is needed to make the measurement.

The next stage is to look at how to test whether a hypothesis is justified or not.

Check your answers at **www.hoddereducation.com/cambridgeextras**

WORKED EXAMPLE 2

It is suggested that the time (t) taken to melt an ice cube is inversely proportional to θ^2, where θ is the initial temperature of the water in °C. Explain whether or not your results from Worked example 1 support this theory.

Answer

If $t \propto \dfrac{1}{\theta^2}$ then $t \times \theta^2 = $ constant

Trial 1: $85 \times 50^2 = 213\,000$

Trial 2: $31 \times 80^2 = 198\,400$

difference between the constants $= 14\,600$

percentage difference $= \dfrac{14\,600}{198\,400} \times 100\% = 7.4\%$

This is less than the calculated uncertainty in the measurement of t ($= 16\%$, for trial 2) so the hypothesis is supported.

There are various ways of tackling this type of problem – this is probably the simplest. Note that it is important to explain fully why the hypothesis is/is not supported. At the simplest level, if the difference between the two calculated values for the constant is greater than the percentage uncertainties in the measured quantities, then the evidence would not support the hypothesis.

STUDY TIP

A more sophisticated approach in this example would be to consider the combined uncertainties in the raw readings as the limit at which the experiment supports the theory. The theory predicts that $t = $ constant$/\theta^2$, which means that the constant $= t \times \theta^2$. To combine uncertainties on multiplication (or division), the percentage uncertainties are added.

» percentage uncertainty in $\theta = 2\%$ (see above, the greatest uncertainty is chosen)
» therefore, percentage uncertainty in $\theta^2 = 2 \times 2\% = 4\%$
» percentage uncertainty in $t = 16\%$
» total uncertainty $= 4\% + 16\% = 20\%$

Identifying limitations

REVISED ☐

There are two parts to this section:

» identifying weaknesses in the procedure
» suggesting improvements that would increase the reliability of the experiment

Before looking at Worked example 3, try to list *four* weaknesses in the procedure in the previous experiment. Then list *four* improvements that would increase the reliability of the experiment.

WORKED EXAMPLE 3

State four sources of error or limitations of the procedure in Investigation 3 – the melting ice experiment.

Answer

1 Two readings are not enough to make firm conclusions.

2 The ice cubes may not be the same mass.
3 There will be some energy exchanges with the surroundings.
4 The ice cubes might be partly melted before they are put into the water.

» Identifying weaknesses in a procedure is not easy, but the more practical work you do, the better you will become.
» It is important to be precise when making your points.
» In many experiments (not this one), parallax can lead to errors. It would not be enough to suggest 'parallax errors' as a limitation. You would need to identify where those errors arose. If you were trying to measure the maximum amplitude of a pendulum, you would need to say, 'Parallax errors, when judging the highest point the pendulum bob reaches'.

» Having identified the areas of weakness, you now need to suggest how they could be rectified. The list given in Worked example 4 is not exhaustive – for example, a suggestion that there should be the same volume of water in the beaker every time would also be sensible. However, a comment regarding measuring the average temperature of the water would not be acceptable as this would make it a different experiment.

If you have not got four weaknesses, try writing 'cures' for the weaknesses suggested in Worked example 3.

WORKED EXAMPLE 4

Suggest four improvements that could be made to Investigation 3. You may suggest the use of other apparatus or different procedures.

Answer

1   Take more sets of readings with the water at different temperatures and plot a graph of t against $1/\theta^2$.

2   Weigh the ice cubes.

3   Carry out the experiment in a vacuum flask.

4   Keep the ice cubes in a cold refrigerator until required.

» In many ways, this is easier than identifying weaknesses but note that you need to make clear what you are doing.
» The first suggestion is a good example – there is no point in taking more readings unless you do something with them.
» Note also that the answer makes it clear that it is not just repeat readings that would be taken (that should have been done anyway); it is readings at different water temperatures.
» This experiment does not cover all the difficulties you might encounter; for instance, in the bouncing ball experiment (Investigation 1), the major difficulty is measuring the height to which the ball bounces. One possible way in which this problem could be solved is to film the experiment and play it back frame by frame or in slow motion.
» Whenever you carry out an experiment, think about the weaknesses in the procedure and how you would rectify them.
» Discuss your ideas with your friends and with your teacher. You will find that you gradually learn the art of critical thinking.

Check your answers at **www.hoddereducation.com/cambridgeextras**

Exam-style questions

This section contains structured questions similar to those you will meet in Paper 2.

In the actual examination, you have 1 hour and 15 minutes to do the paper. There are 60 marks on the paper, so you can spend just over 1 minute per mark. If you find you are spending too long on one question, then move on to another that you can answer more quickly. If you have time at the end, then come back to the difficult one. When you do the exam-style questions in the book, try to stick to this time schedule – it will give you a guide to the rate you need to work at in the examination.

Some questions require you to recall information that you have learned. Be guided by the number of marks awarded to suggest how much detail you should give in your answer. The more marks there are, the more information you need to give.

Some questions require you to use your knowledge and understanding in new situations. Do not be surprised to find something completely new in a question – something you have not seen before. Just think carefully about it, and find something that you do know that will help you to answer it.

Think carefully before you begin to write. The best answers are short and relevant – if you target your answer well, you can get a lot of marks for a small amount of writing. As a general rule, there will be twice as many answer lines as marks. So you should try to answer a 3 mark question in no more than 6 lines of writing. If you are writing much more than that, you almost certainly have not focused your answer tightly enough.

Look carefully at exactly what each question wants you to do. For example, if it asks you to 'Explain', then you need to say *how* or *why* something happens, not just *describe* what happens. Many students lose large numbers of marks by not reading the question carefully.

Where there are calculations to be done, write out the formula and show the examiner what you are doing. There are often several marks for the calculation and even if a simple error, such as an arithmetic error, occurs early in the calculation, the examiner can still give you marks for the rest of the calculation, as long as they can understand what you are doing.

Go to **www.hoddereducation.com/cambridgeextras** for sample answers and commentaries to the questions.

For each question in this practice paper, there is an answer that might get a C or D grade, followed by expert comments (shown by the icon 🄴). Then there is an answer that might get an A or B grade, again followed by expert comments. Try to answer the questions yourself first, before looking at the answers and comments online.

Chapter 1

1 The frequency f of a stationary wave on a string is given by the formula:

$$f = \frac{\sqrt{T/\mu}}{2L}$$

where L is the length of the string, T is the tension in the string and μ is a constant for the string.

a State which of the quantities f, L and T are base quantities. [1]

b i State the SI base units of T. [1]

ii Determine the SI base units of μ. [1]

[Total: 3]

2 a Describe the difference between a scalar quantity and a vector quantity. [2]

b A simple pendulum has a bob of mass 50 g. Determine the weight of the bob. [2]

c The bob is pulled to one side by a horizontal force so that the pendulum string makes an angle of 30° with the vertical.

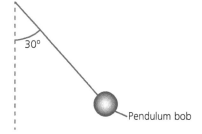

i Add arrows to the diagram to show the direction of:

1 the weight of the pendulum bob (label this *W*)

2 the horizontal force (label this *F*)

3 the tension (label this *T*) [3]

ii Calculate the magnitude of the tension in the string. [2]

iii Calculate the magnitude of the horizontal force. [2]

[Total: 11]

Chapter 2

1 A builder throws a brick up to a second builder on a scaffold, who catches it. The graph shows the velocity of the brick from when it leaves the hand of the first builder to when the second builder catches it.

Velocity/ m s⁻¹ vs **Time/s** graph

  a Show that the acceleration of the brick is $9.8\,\mathrm{m\,s^{-2}}$. [2]

  b The gradient of the velocity–time graph is negative.
     Explain what this shows. [1]

  c The second builder catches the brick 1.04 s after the first builder released it.
     Calculate the height the second builder is above the first builder. [2]

  d The second builder drops a brick for the builder on the ground to catch.
     Suggest why it is much more difficult to catch this brick than the one in the previous case. [1]

[Total: 6]

2 A remote controlled vehicle on the planet Mars fires an object at a velocity of $25\,\mathrm{m\,s^{-1}}$. The initial direction of the velocity is at 60° to the surface of Mars. You may assume that the air resistance on Mars is negligible.
  (acceleration of free fall on Mars = $3.7\,\mathrm{m\,s^{-2}}$)

  a Sketch the flight path of the object. [1]

  b Calculate:
     i   the vertical component of the object's initial velocity [1]
     ii the horizontal component of the object's velocity [1]

  c Show that the time of flight of the object before it hits the ground is 12 s. [3]

  d Calculate the horizontal distance that the object travels before landing on the ground. [2]

  e Explain why the object would travel further when fired on Mars compared with firing it with the same velocity on Earth. [2]

[Total: 10]

Chapter 3

1 a Explain the difference between mass and weight. [2]

  b A golf ball of mass 46 g travelling horizontally collides with a vertical brick wall at 90°. The speed of the ball just before it collides with the wall is $25\,\mathrm{m\,s^{-1}}$. It rebounds horizontally from the wall at a speed of $23\,\mathrm{m\,s^{-1}}$.
     Calculate the change in momentum of the ball. [3]

  c The ball is in contact with the wall for $4.0 \times 10^{-2}\,\mathrm{s}$.
     Calculate the average force acting on the ball during the collision. [2]

  d Sketch the path the ball will take after it hits the wall. [1]

[Total: 8]

2 Glider A on an air track has a mass of 1.2 kg. It moves at $6.0\,\mathrm{m\,s^{-1}}$ towards glider B which is stationary and has a of mass 4.8 kg (Figure 3.8).

Glider A 1.2 kg   Glider B 4.8 kg

The two gliders collide and glider A rebounds with a speed of $3.6\,\mathrm{m\,s^{-1}}$.

  a State what is meant by an elastic collision. [1]

  b Show that the speed of the second glider after the collision is $2.4\,\mathrm{m\,s^{-1}}$. [2]

  c Show that the collision is elastic. [3]

The gliders are in contact for 30 ms during the collision.

  d i   Calculate the average force on the stationary glider during the collision. [2]
     ii Compare the forces on the two gliders during the collision. [2]

[Total: 10]

Chapter 4

1 a State the conditions for an object to be in equilibrium. [2]

  b A non-uniform trap door of mass 6.8 kg is held open by a force of magnitude 24 N applied perpendicular to the door. The trap door is 1.8 m long.
     Calculate the distance from the hinge to the centre of gravity. [3]

Check your answers at **www.hoddereducation.com/cambridgeextras**

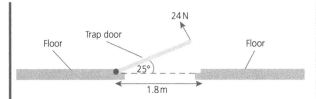

c As well as the two forces shown in the diagram, there is a reaction force.
Use a triangle of forces to show the direction in which this force acts. [3]

[Total: 8]

2 a Explain what is meant by the *moment of a force* about a point. [2]

The diagram shows the principle of a hydraulic jack. A vertical force is applied at one end of the lever, which is pivoted at A. The plunger is pushed down, creating a pressure on the oil, which is pushed out of the master cylinder, through the valve into the slave cylinder.

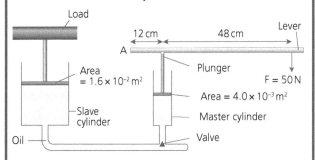

b i Calculate the force produced on the plunger by the lever. [2]

ii Calculate the pressure exerted on the oil by the plunger. [2]

The pressure is transmitted through the oil so that the same pressure is exerted in the slave cylinder.

c Calculate the load the jack can support. [1]

d Suggest two design changes to the jack so that a larger load could be lifted. [2]

[Total: 9]

Chapter 5

1 a State what is meant by the term *work*. [2]

b A car of mass 1200 kg is travelling at a constant speed of 50 km h⁻¹ up an incline at 20° to the horizontal. The engine applies a constant driving force of 5000 N. The engine works with an efficiency of 18%.

i Calculate the total power output from the engine. [2]

ii Determine the input power to the engine. [2]

iii Calculate the fraction of the power output that is used to do work against gravity. [2]

[Total: 8]

2 a i Explain what is meant by the term *energy*. [1]

ii State the principle of conservation of energy. [1]

b A cyclist and her bicycle have a total mass of 82 kg. She is travelling along a straight, level road at a steady speed of 9.0 m s⁻¹. She brakes and her speed drops to 6.4 m s⁻¹.

i Calculate the change in the kinetic energy of the cyclist and her bicycle. [3]

ii State the work done against friction when the bicycle decelerates. [1]

iii Into what form of energy is kinetic energy transferred during the braking process? [1]

[Total: 7]

Chapter 6

1 A projectile launcher uses a compressed spring to provide energy to fire a plastic ball. The compression of the spring and the angle of launch can be adjusted to change the range of the ball. The landing area is a tray of sand with the surface of the sand at the same height as the base of the ball at launch.

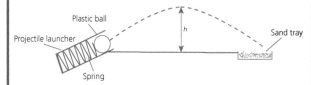

A force of 10 N applied to the spring causes it to compress by 4.0 cm. The mass of the ball is 4.0 g and the angle between the launcher and the horizontal is 24°.

a i Calculate the spring constant for the spring. [2]

ii Calculate the elastic potential energy of the spring when it is compressed by 7.0 cm. [2]

b When the spring is released, all the elastic potential energy is converted to kinetic energy of the ball.

i Calculate the velocity of the ball as it leaves the spring. [2]

ii Calculate the vertical component of the velocity. [2]

iii Determine the maximum height reached by the ball. [2]

[Total: 10]

2 a i Define tensile stress. [1]

ii Define tensile strain. [1]

b The apparatus shown in the diagram is used to determine the Young modulus of a copper wire. The diameter of the wire is 0.193 mm and the original length of the wire is 2.40 m. Masses are added to the wire and the resulting extension is measured using the vernier scale.

The results are shown in the table.

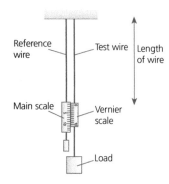

Mass added/kg	Extension of wire/mm	Stress/Pa	Strain
0	0.0		
0.050	0.5		
0.100	0.9		
0.150	1.4		
0.200	1.8		
0.250	2.3		
0.300	2.8		

i Copy and complete the table. [2]

ii Plot a graph of stress against strain. [4]

iii Determine the gradient of the graph. [2]

iv Calculate the value of the Young modulus of the copper wire (with the correct unit). [1]

c The Young modulus is only valid when the wire is deformed within the limit of proportionality.

i Explain what is meant by the limit of proportionality for a wire. [1]

ii Describe how the experimenter could ensure that the wire does not exceed the limit of proportionality during the experiment. [2]

[Total: 14]

Chapter 7

1 The diagram shows a progressive wave moving along a spring at a speed of 3.2 m s^{-1}.

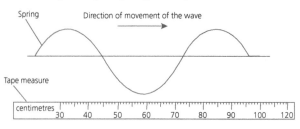

a i On a copy of the diagram, mark the amplitude of the wave. [1]

ii Determine the wavelength of the wave. [1]

iii Calculate the frequency of the wave. Include a unit. [3]

b A police car has a siren that emits a constant sound of frequency 640 Hz. It is travelling along a straight road at a steady speed. A stationary observer on the side of the road detects the sound as having a frequency of 600 Hz.

The speed of sound in air is 330 m s^{-1}.

i Determine the direction of the car's velocity relative to the observer. [1]

ii Calculate the speed of the car. [3]

[Total: 9]

2 a Describe the difference between transverse waves and longitudinal waves. [2]

Light from the Sun is unpolarised, whereas sunlight reflected from some surfaces is partially polarised in a horizontal plane. Polarising spectacles are used to reduce glare when driving.

b i Explain the difference between unpolarised and plane-polarised light. [2]

ii Explain why the polarising spectacles reduce glare but do not totally eliminate it. [2]

c Light from a filament lamp is incident successively on two polarising filters. The second filter's transmission axis is at an angle of 60° to the transmission axis of the first filter.

Calculate the percentage change in the intensity of the transmitted light compared with the light incident on the first filter.

(You may assume that the filters are close together, so that there is negligible spreading of the light between entering the first filter and leaving the second filter.) [3]

[Total: 9]

Check your answers at **www.hoddereducation.com/cambridgeextras**

Chapter 8

1 a State **two** differences between a stationary wave and a progressive wave. [2]

   b A small loudspeaker is connected to a variable frequency signal generator. The loudspeaker is sounded just above a closed air column of length 0.60 m. The frequency from the signal generator is 128 Hz when the first resonance is heard.

      i Calculate the wavelength of the sound wave. [2]

      ii Calculate the speed of the sound in the column. [2]

      iii Calculate the wavelength of the 4th harmonic for a column of this length. [1]

[Total: 7]

2 a i Explain what is meant when two sources of light are described as *coherent* and state the conditions necessary for coherence. [2]

      ii Explain why two separate light sources cannot be used to demonstrate interference of light. [2]

   b A Young's double-slit experiment is set up to measure the wavelength of red light.

      i The slit separation is 1.2 mm and the screen is 3.0 m from the slits. The diagram shows the interference pattern that is observed. Calculate the wavelength of the light. [2]

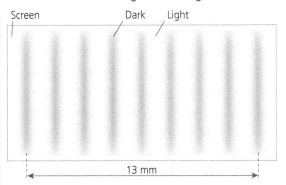

Screen   Dark   Light

13 mm

      ii Explain how you would expect the pattern to change if the red light was replaced by blue light. [2]

[Total: 8]

Chapter 9

1 a Explain what is meant by the term *number density of charge carriers*. [1]

   b A silver wire of radius 0.24 mm carries a current of 0.30 A. The electron number density in silver is 5.9×10^{28} m^{-3}.

      i Calculate the number of electrons flowing past a point in the wire in 1 minute. [2]

      ii Calculate the drift velocity of the electrons. [3]

   c The wire has a resistance of 0.45 Ω. Calculate the power dissipated in the wire. [2]

[Total: 8]

2 a State Ohm's law. [2]

   b A length of Eureka wire has a resistance of 4.0 Ω.

      A student connects the Eureka wire across the terminals of a cell. The potential difference across the wire is 1.56 V. Calculate:

      i the current flowing through the wire [2]

      ii the power dissipated by the wire [2]

   c The Eureka wire is replaced with one of the same length and half the diameter.

      i Determine the resistance of this Eureka wire. [2]

      ii Compare the power dissipated in this wire with the power dissipated in the wire in part b. [2]

[Total: 10]

Chapter 10

1 a Define electromotive force. [2]

   b A student connects a cell, an ammeter, a switch and a resistor in series. A voltmeter is connected in parallel with the cell.

      Draw a circuit diagram of this circuit. [2]

   c When the switch is open, the voltmeter reads 1.54 V. When the switch is closed the voltmeter reading drops to 1.28 V.

      Explain why the voltmeter reading falls when the switch is closed. [2]

   d When the switch is closed, the ammeter reading is 0.40 A.

      Calculate:

      i the resistance of the resistor [2]

      ii the internal resistance of the cell [2]

[Total: 10]

2 a State Kirchoff's second law. [2]
  b Use Kirchoff's second law to derive a formula for the combined resistance of two resistors in series. [2]
  c A battery is connected in series with an ammeter, a resistor and a thermistor. A voltmeter is connected across the resistor. The thermistor is immersed in iced water at 0°C. The ice melts and the temperature of the water rises to 20°C.
   As the temperature of the water rises:
   i Describe what happens to the resistance of the thermistor. [1]
   ii Explain what happens to the p.d. across the resistor. [2]
   iii State and explain whether the ammeter reading rises, falls or remains the same. [2]

[Total: 9]

Chapter 11

1 The α-particle scattering experiment provides evidence about the structure of atoms. The main results were:
» most of the alpha particles passed through the foil with very little deviation
» a small number of alpha particles were deflected through an angle of more than 10°
» very few particles were deflected through an angle greater than 90°.
  a With reference to these results, suggest what conclusions can be drawn about the size and mass of the nucleus. [3]
  b Uranium-235 decays by alpha decay to thorium. Complete the decay equation to show the mass and nucleon numbers of the decay products. [4]

$$^{235}_{92}U \rightarrow Th + \text{———}$$

[Total: 7]

2 The isotope $^{228}_{91}Pa$ decays by positron emission to form an isotope of thorium. The decay can be represented as:

$$^{228}_{91}Pa \rightarrow ^{228}_{90}Th + ^{0}_{+1}\beta$$

  a For the isotope of thorium:
   i Calculate how many neutrons are in the nucleus. [1]
   ii Calculate the mass of the nucleus in atomic mass units. [2]
  b The positron is produced when a proton in the protactinium nucleus changes into a neutron.
   i Describe the quark composition of a proton. [1]
   ii Describe the changes that take place in the quark composition of a proton during β+ decay. [2]
   iii A β+ particle is an example of a lepton. Give the name of one other lepton produced during this decay. [1]
  c i Describe the difference between mesons and baryons. [2]
   ii A certain meson consists of one up quark and one antistrange quark. Determine the relative charge of this meson. Justify your answer by reference to the charge on each quark. [2]

[Total: 11]

Check your answers at **www.hoddereducation.com/cambridgeextras**

12 Motion in a circle

Radian measurement

You are familiar with the use of degrees to measure angles, with a complete circle equal to 360°. There is no real reason why a circle is split into 360° – it probably arises from the approximate number of days it takes for the Earth to orbit the Sun (Figure 12.1).

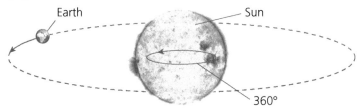

▲ Figure 12.1

It is much more convenient to use **radians**.

$$\text{angle (in radians)} = \frac{\text{arc length}}{\text{radius}}$$

For a complete circle, the circumference = $2\pi r$, where r is the radius. Hence, the angle subtended by a complete circle is:

$$360° = \frac{2\pi r}{r} = 2\pi \text{ radians}$$

This can be expressed as:

$$1° = \frac{2\pi}{360} \text{ rad}$$

Or:

$$1 \text{ rad} = \frac{360}{2\pi} = 57.3°$$

> **KEY TERMS**
>
> One **radian** is the angle subtended at the centre of a circle by an arc of equal length to the radius of the circle (Figure 12.2).
>
>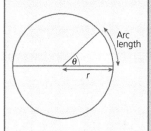
>
> **▲ Figure 12.2** When arc length = r, θ = 1 radian

WORKED EXAMPLE

a Convert the following angles to radians:
  i 180°   ii 60°

b Convert the following angles to degrees:

  i $\frac{\pi}{4}$ rad   ii $\frac{2\pi}{3}$ rad

Answer

a i $180° = 180 \times \frac{2\pi}{360} \text{ rad} = \pi \text{ rad}$ ii $60° = 60 \times \frac{2\pi}{360} \text{ rad} = \frac{\pi}{3} \text{ rad}$

b i $\frac{\pi}{4} \text{ rad} = \frac{\pi}{4} \times \frac{360}{2\pi} = 45°$ ii $\frac{2\pi}{3} \text{ rad} = \frac{2\pi}{3} \times \frac{360}{2\pi} = 120°$

▶ NOW TEST YOURSELF

1 Convert the following angles to degrees:

  a $\frac{\pi}{6}$ rad   b 3π rad   c $\frac{5\pi}{3}$ rad

2 Convert the following angles to radians:

  a 10°   b 48°   c 630°

Angular displacement and angular speed

Consider a particle moving at constant speed (v) round a circle. The change in angle from a particular reference point is called the **angular displacement** (θ) (Figure 12.3).

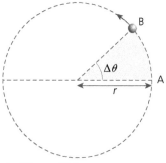

▲ **Figure 12.3**

As the particle moves round the circle, the angular displacement increases at a steady rate. The rate of change in angular displacement is called the **angular speed** (ω).

Comparison with translational motion

Many of the concepts we met in kinematics at AS Level have their equivalent in circular motion. This is shown in Table 12.1.

▼ **Table 12.1**

Translational motion			Circular motion		
Quantity	**Unit**	**Relationship**	**Quantity**	**Unit**	**Relationship**
Displacement (s)	m		Angular displacement (θ)	rad	
Velocity (v)	m s^{-1}	$v = \dfrac{\Delta s}{\Delta t}$	Angular speed (ω)	rad s^{-1}	$\omega = \dfrac{\Delta \theta}{\Delta t}$

Useful equations

Look at Figure 12.3. One way of finding the angular speed (ω) of an object is to measure the time for one complete circuit of the object round a circular track. The time taken is the time period of the revolution (T) and the angular displacement is 2π radians (360°). Therefore:

$$\omega = \frac{2\pi}{T}$$

Again, refer to Figure 12.3. From this we see that:

$$\omega = \frac{\Delta \theta}{\Delta t}$$

But:

$$\Delta \theta = \frac{AB}{r}$$

Therefore:

$$\omega = \frac{AB}{r\Delta t}$$

$$\frac{AB}{\Delta t} = \frac{\text{distance travelled}}{\text{time taken}} = v$$

Thus:

$$\omega = \frac{v}{r}$$

Or, rearranging the formula:

$$v = \omega r$$

KEY TERMS

Angular displacement is the change in angle (measured in radians) of an object as it rotates round a circle.

Angular speed is the change in angular displacement per unit time:

$$\omega = \frac{\Delta \theta}{\Delta t}$$

WORKED EXAMPLE

A car is travelling round a circular bend of radius 24 m at a constant speed of 15 m s^{-1}. Calculate the angular speed of the car.

Answer

$$\omega = \frac{v}{r} = \frac{15}{24} = 0.625 \approx 6.3\,\text{rad s}^{-1}$$

> ## NOW TEST YOURSELF
>
> TESTED ☐
>
> 3 An athlete in the 'throwing the hammer' event whirls a heavy iron ball around his head in a circular path with a radius of 1.2 m. The time taken for the ball to make one complete revolution is 0.80 s. Calculate the angular speed of the ball.
>
> 4 A toy train goes round a circular track. It makes two complete revolutions in 16 s. Calculate the angular speed of the train.
>
> 5 A girl ties a rubber bung to a string. She swings the string so that the bung moves in a circular path. The radius of the circular path is 80 cm and the angular speed of the bung is $4\pi/3$ rad s^{-1}. Calculate the speed at which the bung moves.

Centripetal acceleration and centripetal force

Constant speed, constant acceleration

REVISED ☐

You have seen how an object can move at constant speed round a circle, but what is meant when it is said that the object has a constant acceleration? To understand this, you must remember the definition of acceleration: the change in velocity per unit time. Velocity, unlike speed, is a vector and so a change in direction is an acceleration (Figure 12.4).

(a)

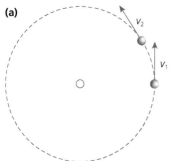

(b)

▲ Figure 12.4

» Consider a particle moving round a circle, as in Figure 12.4(a).
» At time t it has a velocity of v_1.
» After a short interval of time, Δt, it has the velocity v_2 – the same magnitude, but the direction has changed.
» Figure 12.4(b) shows the change of velocity Δv (refer back to vector addition, p. 13).
» You can see that this is towards the centre of the circle.
» The acceleration is $\Delta v/\Delta t$.
» As the object moves round the circle, the direction of its velocity is continuously changing.
» However, the direction of the change is always towards the centre of the circle.
» Thus, the particle has an acceleration of constant magnitude but whose direction is always towards the centre of the circle.
» Such an acceleration is called a **centripetal acceleration**.
» The magnitude of the acceleration a is given by:

$$a = \frac{v^2}{r} = \omega^2 r$$

> ### STUDY TIP
>
> Use tracing paper to copy Figure 12.4(b). Move the tracing paper so that v_1 rests on v_1 on Figure 12.4(a).
>
> You should be able to see that Δv points towards the centre of the circle.

Centripetal force and acceleration

» An object travelling round a circle at constant speed is not in equilibrium.
» From Newton's laws, you will remember that for an object to accelerate, a resultant force must act on it.
» The force must be in the same direction as the acceleration.
» Hence, the force is always at right angles to the velocity of the object, towards the centre of the circle (a centripetal force).
» Such a force has no effect on the magnitude of the velocity; it simply changes its direction.

Using the relationship $F = ma$ (where F = force and m = mass of the object), we can see that the force can be calculated from:

$$F = \frac{mv^2}{r} = m\omega^2 r$$

> **STUDY TIP**
>
> It is often thought that an object rotating round a circle at constant speed is in equilibrium – just remember that it continuously changes direction, hence there is always force on it. If that force is removed, the object will no longer move in a circle but in a straight line at a tangent to the original circle.

▶ REVISION ACTIVITY

You should be able to develop many equations from more fundamental equations. Some of these fundamental equations are given at the beginning of the examination paper. Others you must learn by heart. It is a good idea to write out these equations on a piece of card and stick the card on a wall at home to learn them by heart.

'Must learn' equations:

$$\omega = \frac{\Delta\theta}{\Delta t} \qquad a = \frac{v^2}{r} = \omega^2 r$$

$$\omega = \frac{v}{r} \qquad F = \frac{mv^2}{r} = m\omega^2 r$$

Figure 12.5(a) shows a rubber bung being whirled round on a string. The string is under tension.

(a)
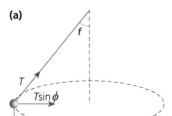

(b)
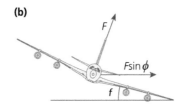

▲ **Figure 12.5**

The centripetal force is the component of the tension in the horizontal direction ($T\sin\phi$).

$$F = \frac{mv^2}{r} = T\sin\phi$$

In Figure 12.5(b), the uplift on the aeroplane is perpendicular to the wings. When the aeroplane banks there is a horizontal component to this, which provides a centripetal force ($F\sin\phi$) and the plane moves along the arc of a circle.

$$\frac{mv^2}{r} = F\sin\phi$$

> **STUDY TIP**
>
> Whenever an object moves at constant speed in a circle, there is a force towards the centre of the circle, known as a centripetal force. What provides this force varies according to the situation.
>
> Example 1: The force when a ball attached to an elastic cord is spun round in a circle is supplied by the tension in the cord.
>
> Example 2: The force when a satellite orbits around Mars is supplied by the gravitational force between Mars and the satellite.
>
> Example 3: The force when a car goes round a level circular track is supplied by friction between the car's tyres and the road.

> ## NOW TEST YOURSELF
> TESTED ☐
>
> 6 Calculate the centripetal acceleration of the toy train in question 4 (p. 117) and the centripetal force on it. The diameter of the track is 1.2 m and the mass of the train is 0.45 kg.
> 7 A car of mass 800 kg goes round a bend at a speed of 15 m s^{-1}. The path of the car can be considered to be an arc of a circle of radius 25 m. Calculate:
>   a   the angular speed of the car
>   b   the centripetal force on the car

WORKED EXAMPLE

Figure 12.6 shows a racing car rounding a bend of radius 120 m on a banked track travelling at 32 m s^{-1}.

a   Calculate the angle ϕ if there is no tendency for the car to move either up or down the track. You may treat the car as a point object.

b   Suggest and explain what would happen if the car's speed was reduced.

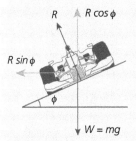

▲ Figure 12.6

Answer

a R is the normal reaction force.
Resolving vertically:

$$R\cos\phi = mg$$

Resolving horizontally:

$$R\sin\phi = \frac{mv^2}{r}$$

Dividing the two equations:

$$\frac{\sin\phi}{\cos\phi} = \frac{mv^2/r}{mg}$$

$$\tan\phi = \frac{v^2}{gr} = \frac{32^2}{9.81 \times 120} = 0.871$$

$$\phi = 41°$$

b   The car would tend to slip down the slope as the required centripetal force would be less. In practice, frictional forces would probably mean that it would continue in a circle of the same radius.

> ## STUDY TIP
>
> The car is clearly not a point object but modelling it as one simplifies the problem. The normal reaction is, in reality, shared at each of the four wheels. The wheels on the outside of the curve travel in a larger circle than those on the inside of the circle, further complicating the picture. Engineers and scientists often use simplified models, which they then develop to solve more complex problems.

> ## REVISION ACTIVITY
>
> A racing car hits a patch of oil as it enters a bend.
>
> Explain why the car slides off the road into the gravel trap.

> ## END OF CHAPTER CHECK
>
> In this chapter, you have learnt to:
> » define the radian ☐
> » express angular displacement in radians ☐
> » understand and use the concept of angular speed ☐
> » recall and use the formulae $\omega = 2\pi/T$ and $v = \omega r$ ☐
> » understand that a force of constant magnitude that is always perpendicular to the direction of motion causes centripetal acceleration ☐
>
> » understand that centripetal acceleration causes circular motion with a constant angular speed ☐
> » recall and use $a = \omega^2 r$ ☐
> » recall and use $a = v^2/r$ ☐
> » recall and use $F = m\omega^2 r$ ☐
> » recall and use $F = mv^2/r$ ☐

13 Gravitational fields

Gravitational field

A **gravitational field** is a region around an object that has mass, in which another object with mass experiences a force.

» A gravitational field is an example of a field of force.
» Fields of force can be represented by lines of force.
» A line of gravitational force shows the direction of the force an object placed at a point in the field experiences.
» The closer the lines of force in a field diagram, the stronger the field.

Gravitational field strength

Any object near the Earth's surface is attracted towards the Earth with a force that is dependent on the mass of the object. Similarly, an object near the Moon is attracted towards the Moon's surface, but the force is smaller. The reason for this is that the **gravitational field strength** is greater near the Earth than it is near the Moon.

gravitational force on an object = mass of the object × gravitational field strength

In symbols:

$$F = mg$$

You might remember g as the acceleration due to gravity or acceleration of free fall (p. 27), but if you compare the formulae $F = ma$ and $F = mg$, you can see that the acceleration due to gravity and the gravitational field strength are the same thing.

The lines of force in Figure 13.1 are extremely near to being parallel and they are all equal distances apart. This shows that the gravitational field strength near the Earth's surface is (very nearly) uniform.

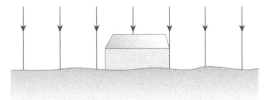

▲ Figure 13.1

> **KEY TERMS**
>
> The **gravitational field strength** at a point is defined as the gravitational force per unit mass at that point. The units of gravitational field strength are $N\,kg^{-1}$.

Gravitational force between point masses

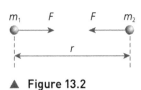

▲ Figure 13.2

It is not just large objects that attract each other – all masses have a gravitational field. This means that they attract other masses (Figure 13.2).

Two point masses of mass m_1 and m_2 separated by a distance r will attract each other with the magnitude of the force given by the formula:

$$F = G\frac{m_1 m_2}{r^2}$$

» G is a constant known as the universal gravitational constant. Its value is $6.67 \times 10^{-11}\,N\,m^2\,kg^{-2}$.
» This is known as Newton's law of gravitation.

Check your answers at **www.hoddereducation.com/cambridgeextras**

> **NOW TEST YOURSELF**   TESTED ☐

1 Two point masses, each of magnitude 5.0 kg, are placed 15 cm apart. Calculate the force on each object.

Objects of finite size   REVISED ☐

>> It is slightly more complex with objects of finite size.
>> All the mass of any object can be considered to act at a single point, which is called the **centre of mass**.
>> For a uniform sphere, the centre of mass is at the centre of the sphere. The Earth, the Sun, the Moon and other planets may be considered to be uniform spheres.
>> This simplifies the maths and, in effect, the object is treated as a point mass.
>> However, you must be careful to remember to measure any distances between objects as the distance between their centres of mass, *not* between their surfaces.

WORKED EXAMPLE

Two spheres of radius 0.50 cm and masses 150 g and 350 g are placed so that their centres are 4.8 cm apart.
a Calculate the force on the 150 g sphere.
b Write down the force on the 350 g sphere.

Answer

a 350 g = 0.35 kg, 150 g = 0.15 kg, 4.8 cm = 0.048 m

$$F = G\frac{m_1 m_2}{r^2}$$
$$= 6.67 \times 10^{-11} \times \frac{0.35 \times 0.15}{0.048^2}$$
$$= 1.5 \times 10^{-9}\,\text{N}$$

b In accordance with Newton's third law, the force on the 350 g mass will also be 1.5×10^{-9} N but in the opposite direction.

> **STUDY TIP**
>
> This shows how small the gravitational attraction between two small objects is. It is only when we consider planet-sized objects that the forces become significant.

> **NOW TEST YOURSELF**   TESTED ☐

2 Calculate the gravitational attraction between the Earth and the Moon. (mass of the Earth = 6.0×10^{24} kg, mass of the Moon = 7.3×10^{22} kg, separation of the Earth and Moon = 3.8×10^8 m)

Gravitational field of a point mass

>> The gravitational field strength has already been defined as the gravitational force per unit mass at that point.
>> In earlier work, you only considered gravitational fields on large objects such as the Earth and other planets, and then only near their surfaces.
>> Under these circumstances, the field may be considered uniform.
>> However, the gravitational field of a point mass is radial (Figure 13.3a).
>> This is also true for any object of finite size, as long as we are outside the object itself.
>> In the case of an object of finite size, the radial field is centred on the centre of mass of the object (Figure 13.3b).

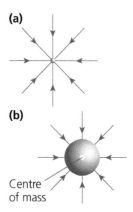

(a)

(b)

Centre of mass

▲ Figure 13.3 The gravitational field of (a) a point mass and (b) an object of finite size

You can see from Figure 13.3 that the lines of gravitational force get further apart as the distance from the centre of mass increases. This shows that the field strength decreases with increasing distance from the object.

Consider the equation for the magnitude of the gravitational force between two objects and the definition of gravitational field strength:

$$F = G\frac{Mm}{r^2} \text{ and } g = \frac{F}{m}$$

$$g = \frac{GM}{r^2}$$

The equation shows an **inverse square** relationship (Figure 13.4). This means if the distance from the mass is doubled, the field decreases by a factor of 4 (2^2).

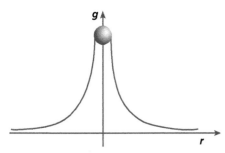

▲ Figure 13.4 The gravitational field near a spherical object

WORKED EXAMPLE

Calculate the gravitational field strength at the surface of Mars.
(radius of Mars = 3.4×10^3 km, mass of Mars = 6.4×10^{23} kg, $G = 6.67 \times 10^{-11}$ N m^2 kg^{-2})

Answer

$$3.4 \times 10^3 \text{ km} = 3.4 \times 10^6 \text{ m}$$

$$F = G\frac{Mm}{r^2} \text{ and } g = \frac{F}{m}$$

$$g = \frac{GM}{r^2} = \frac{(6.67 \times 10^{-11}) \times (6.4 \times 10^{23})}{(3.4 \times 10^6)^2} = -3.7 \, \text{N kg}^{-1}$$

The magnitude of the gravitational field strength is 3.7 N kg^{-1} towards the centre of Mars.

Orbital mechanics

REVISED ☐

Figure 13.5 shows a satellite travelling in a circular orbit around the Earth.

The gravitational pull on the satellite provides the centripetal force to keep the satellite in orbit.

Centripetal force:

$$F = \frac{mv^2}{r} = G\frac{Mm}{r^2}$$

Cancelling m and r:

$$v^2 = \frac{GM}{r}$$

which can be rewritten as:

$$v = \sqrt{\frac{GM}{r}}$$

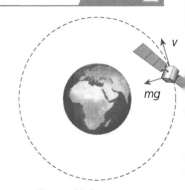

▲ Figure 13.5

This can also be expressed in terms of angular speed, ω:

$$v = \omega r$$

Check your answers at **www.hoddereducation.com/cambridgeextras**

Therefore:

$$\omega = \sqrt{\frac{GM}{r^3}}$$

You can see that the angular speed, and hence the frequency and the period for one orbit, are dependent on the orbital radius.

The relation between the period T for one orbit and the angular speed ω is:

$$T = \frac{2\pi}{\omega}$$

and between the frequency f and the period it is:

$$f = \frac{1}{T}$$

WORKED EXAMPLE

The International Space Station (ISS) orbits 400 km above the Earth's surface. Calculate:

a the period of the orbit

b the speed of the ISS

(mass of Earth = 6.0×10^{24} kg, radius of Earth = 6.4×10^3 km)

Answer

a orbital radius of the ISS = Earth's radius + height of the ISS above the surface

$$= (6.4 \times 10^3 + 400) = 6.8 \times 10^3 \, \text{km} = 6.8 \times 10^6 \, \text{m}$$

$$\omega = \sqrt{\frac{GM}{r^3}} = \sqrt{\frac{(6.67 \times 10^{-11}) \times (6.0 \times 10^{24})}{(6.8 \times 10^6)^3}} = 1.13 \times 10^{-3} \, \text{rad s}^{-1}$$

$$T = \frac{2\pi}{\omega} = \frac{2\pi}{1.13 \times 10^{-3}} = 5.57 \times 10^3 \, \text{s} = 1.5 \, \text{h}$$

b $v = \omega r$

orbital radius = 6.8×10^3 km

$$v = (1.13 \times 10^{-3}) \times (6.8 \times 10^3) = 7.7 \, \text{km s}^{-1}$$

Geostationary orbits

» A satellite orbits the Earth directly above the equator.
» If the satellite orbits in the same direction as the Earth spins and has an orbital period of 24 hours, it will remain over the same point above the Earth's surface.
» This type of orbit is used for communication satellites (Figure 13.6).

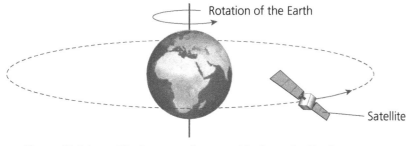

▲ **Figure 13.6 A satellite in geostationary orbit above the Earth**

WORKED EXAMPLE

Calculate the height above the Earth that a satellite must be placed for it to orbit in a geostationary manner.

(mass of Earth = 6.0×10^{24} kg, radius of Earth = 6.4×10^6 m)

Answer

time period required for a geostationary orbit is 24 h = 86 400 s

$$\omega = \frac{2\pi}{T}$$

$$\omega = \sqrt{\frac{GM}{r^3}}$$

So $\frac{2\pi}{T} = \sqrt{\frac{GM}{r^3}}$

And:

$$r^3 = \frac{GMT^2}{(2\pi)^2} = \frac{(6.67 \times 10^{-11}) \times (6.0 \times 10^{24}) \times 86\,400^2}{4\pi^2} = 7.57 \times 10^{22}$$

$$r = \sqrt[3]{7.57 \times 10^{22}} = 4.23 \times 10^7$$

This is the radius of the satellite's orbit. The radius of the Earth is 6.4×10^6 m, so the height of the satellite above the Earth's surface is:

$$(42.3 \times 10^6) - (6.4 \times 10^6) = 3.59 \times 10^7 \text{ m} \approx 3.6 \times 10^7 \text{ m}$$

NOW TEST YOURSELF

TESTED ▢

3 During the Moon landings in the 1970s, the command module orbited the Moon at an orbital height of 130 km above the surface of the moon. (mass of the Moon = 7.3×10^{22} kg, the radius of the Moon = 1.740×10^3 km) Calculate:
   a the period of the orbit
   b the speed at which the satellite moved relative to the Moon's surface

---

Gravitational potential

Gravitational potential energy at a point

REVISED ▢

From earlier work (see p. 47), you will be familiar with the idea that the gain in gravitational potential energy of an object when it is lifted through a height Δh is given by the formula:

$$\Delta E_p = mg\Delta h$$

➤➤ This formula only works if the gravitational force is constant, which is only true if the field is uniform.
➤➤ If the field is non-uniform, it is approximately true for very small changes in height, so a series of tiny changes can be added to give the total change in gravitational potential energy.
➤➤ Physicists define the point at which an object has zero gravitational potential energy as infinity.
➤➤ This means that the gravitational potential energy at an infinite distance from any object is zero.
➤➤ This is a little difficult to start with; we know that an object loses gravitational potential energy as it approaches the Earth or other large object.
➤➤ Thus, it has less than zero gravitational potential energy as it approaches an object.
➤➤ Put another way, it has negative gravitational potential energy when it is near another object such as the Earth.

Check your answers at **www.hoddereducation.com/cambridgeextras**

Gravitational potential

» By considering the gravitational potential energy of a unit mass, we can assign each point in space a specific **gravitational potential** (ϕ).
» Figure 13.7 shows that the gravitational potential at the surface of the object is negative, and how the potential increases towards zero as we move away from the object.

> **KEY TERMS**
>
> The **gravitational potential** at a point is the work done per unit mass in bringing a small test mass from infinity to that point.
>
> The units of gravitational potential are $J\,kg^{-1}$.

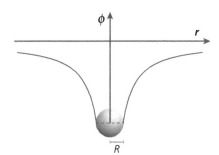

▲ **Figure 13.7 The gravitational potential near an object of radius** R

Solving problems

A careful study of the potential curve shows it to be of the form $\phi \propto 1/r$.

The formula for calculating the gravitational potential at a point is:

$$\phi = -\frac{GM}{r}$$

where r is the distance from the centre of mass of the object.

WORKED EXAMPLE

If an object is fired from the Earth's surface with sufficient speed, it can escape from the Earth's gravitational field.

a Calculate the potential at the Earth's surface.
b State and explain the minimum energy an object of unit mass would need to be given to escape from the Earth's gravitational field.
c Calculate the minimum speed at which the object must be fired to escape.

(Mass of Earth = 6.0×10^{24} kg, radius of Earth = 6.4×10^{6} m)

Answer

a $\phi = -\dfrac{GM}{r} = -\dfrac{(-6.67 \times 10^{-11}) \times (6.0 \times 10^{24})}{6.4 \times 10^{6}} = -6.25 \times 10^{7}\,J\,kg^{-1}$

b 6.25×10^{7} J, the energy required to reach infinity, zero potential energy
c $E_k = \frac{1}{2}mv^2$
which leads to:

$$v = \sqrt{\frac{2E_k}{m}} = \sqrt{\frac{2 \times (6.25 \times 10^{7})}{1}} = 1.1 \times 10^{4}\,m\,s^{-1}$$

> ▶ **NOW TEST YOURSELF**
>
>
> 4 Why was infinity chosen as the zero of potential, rather than the surface of the Earth?
>   Hint: think on the very large scale rather than on the smaller scale.
> 5 Calculate the gravitational potential at the Moon's surface.
>   (mass of the Moon = 7.3×10^{22} kg, radius of the Moon = 1.74×10^{6} m)
> 6 Use your answer to question 5 to calculate the escape velocity from the Moon's surface.

Gravitational potential energy

From the definition of gravitational gravitational potential, it follows that the gravitational potential energy of two isolated point masses is given by the equation:

$$E_p = -\frac{GMm}{r}$$

Very often we consider a small mass, such as a spacecraft or meteorite, near a much larger mass, such as a planet. In those circumstances, we often think of the small mass as having a gravitational potential energy given by the formula $E_p = -GMm/r$ and forget that the planet also has the same magnitude of gravitational potential energy due to the small object.

▶ NOW TEST YOURSELF

TESTED ☐

7 a Estimate your gravitational potential energy when you are standing on the Earth's surface.

b Write down the Earth's gravitational potential energy due to your presence. Hint: the answer is not zero – why not?
(mass of the Earth = 6.0×10^{24} kg, radius of the Earth = 6.4×10^6 m)

▶ REVISION ACTIVITIES

Use the internet to find the orbital period of the International Space Station. Use the information to find the height above the Earth at which it orbits. Check this figure from another internet source.

'Must learn' equations:

$$g = \frac{GM}{r^2} \qquad\qquad \phi = \frac{-GM}{r}$$

$$F = \frac{Gm_1m_2}{r^2} \qquad\qquad E_p = \frac{-m_1m_2}{r}$$

STUDY TIP

Question 7 does not satisfactorily fulfil the criteria for two isolated spherical masses for a variety of reasons. However, we can use this as a model to give a good estimate as to your gravitational potential energy.

▶ END OF CHAPTER CHECK

In this chapter, you have learnt:
- ⟫ that a gravitational field is an example of a field of force ☐
- ⟫ that gravitational field strength is defined as the force per unit mass ☐
- ⟫ to represent a gravitational field by means of field lines ☐
- ⟫ to understand that for a point outside a sphere, the mass of the sphere may be considered to be a point mass at its centre ☐
- ⟫ to recall and use Newton's law of gravitation $F = Gm_1m_2/r^2$ for the force between two point masses ☐
- ⟫ to derive, from Newton's law of gravitation and the definition of gravitational field, the equation $g = Gm/r^2$ for the gravitational field strength due to a point mass ☐
- ⟫ to recall and use $g = Gm/r^2$

- ⟫ to analyse circular orbits in gravitational fields by relating the gravitational force to the centripetal acceleration it causes ☐
- ⟫ to understand that a satellite in a geostationary orbit remains at the same point above the Earth's surface, with an orbital period of 24 hours, orbiting from west to east directly above the Earth's equator ☐
- ⟫ to understand why g is approximately constant for small changes in height near the Earth's surface ☐
- ⟫ to define gravitational potential at a point as the work done per unit mass in bringing a small test mass from infinity to the point ☐
- ⟫ to use $\phi = -GM/r$ for the gravitational potential in the field for a point mass ☐
- ⟫ to understand how the concept of gravitational potential leads to the gravitational potential energy of two masses and use $E_p = -Gm_1m_2/r$ ☐

Check your answers at **www.hoddereducation.com/cambridgeextras**

Thermal equilibrium

What is temperature?

REVISED

» Temperature tells us the direction in which there will be a net energy flow between objects in thermal contact.
» Thermal energy will tend to flow from an object at high temperature to an object at a lower temperature.
» If there is no net energy flow between two objects in thermal contact, then those two objects are at the same temperature. They are said to be in thermal equilibrium.
» Figure 14.1(a) shows that if object A is at a higher temperature than object B, and if object B is at a higher temperature than object C, then object A is at a higher temperature than object C.
» Figure 14.1(b) shows that if object P is in thermal equilibrium with object Q, and if object Q is in thermal equilibrium with object R, then object P is in thermal equilibrium with object R.

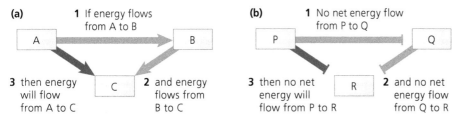

▲ **Figure 14.1 The energy flow between different objects in thermal contact**

> **NOW TEST YOURSELF**   TESTED

1   Consider three objects X, Y and Z in thermal contact. Which of the following is possible?
    A   Energy flows from X to Y, no net energy flow between Y and Z, energy flows from Z to X.
    B   Energy flows from X to Y, no net energy flow between Y and Z, energy flows from X to Z.

C   No net energy flow between X and Y, no net energy flow from Y to Z, energy flows from Z to X.
D   No net energy flow between X and Y, energy flows from Y to Z, energy flows from Z to X.

Temperature scales

Measurement of temperature

REVISED

To measure temperature, a physical property that varies with temperature is used. Examples are:

» expansion of a liquid (including resultant change in density)
» expansion of a gas at constant pressure
» change of pressure of a gas at constant volume
» change in resistance of a metal or a semiconductor
» e.m.f. produced across the junctions of a thermocouple

The thermodynamic, or kelvin, temperature scale

REVISED

»» Temperature scales require two fixed points that are easily repeatable.
»» For example, the Celsius scale uses the melting point of pure water as the lower fixed point (0°C) and the boiling point of pure water at standard atmospheric pressure as the higher fixed point (100°C). When a thermometer is calibrated, these two points are marked and then the scale is divided into 100 equal parts.

The **thermodynamic** temperature scale is different and is independent of the physical properties of any particular substance.

»» The zero on the thermodynamic temperature scale is known as **absolute zero**, the temperature at which no more energy can be removed from any object. All the energy that can be removed has been removed. At this temperature, all substances have minimum internal energy. This does not mean that the object has zero energy. Only an ideal gas (see p. 133) will have zero energy; it will also have zero pressure.
»» Absolute zero is equivalent to −273.15°C.
»» The unit in the thermodynamic scale is the kelvin, symbol K.
»» The magnitude of the kelvin is set by fixing the numerical value of the Boltzmann constant to a predetermined value – see p. 134.

> **STUDY TIP**
> When the thermodynamic temperature scale was first introduced, the size of the unit in the thermodynamic scale was chosen to be the same size as the degree in the old centigrade scale, now redefined as the Celsius scale.

Conversion between Celsius and kelvin scales

REVISED

To convert between kelvin and Celsius we use:

$$T/\text{K} = \theta/°\text{C} + 273.15$$

In practice, we often simplify the conversion by using $T/\text{K} = \theta/°\text{C} + 273$.

WORKED EXAMPLE
Copy and complete Table 14.1, showing your working.

▼ Table 14.1

	Temperature/K	Temperature/°C
Boiling point of water		100
Boiling point of bromine	332.40	
Boiling point of helium	4.37	
Triple point of hydrogen		−259.34
Boiling point of nitrogen	77.50	

Answer

▼ Table 14.2

	Temperature/K	Temperature/°C
Boiling point of water	100 + 273 = 373	100
Boiling point of bromine	332.40	332.40 − 273.15 = +59.25
Boiling point of helium	4.37	4.37 − 273.15 = −268.78
Triple point of hydrogen	−259.34 + 273.15 = 13.81	−259.34
Boiling point of nitrogen	77.50	77.50 − 273.15 = −195.65

> **STUDY TIP**
> The boiling point of water is given only to the nearest degree Celsius. Therefore, using 273 as the difference between Celsius and kelvin is justified.

2 A voltmeter connected to a thermocouple reads 0 V when both junctions are in ice at 0°C, and 4.8 mV when one junction is in ice and one is in boiling water at 100°C.
  What is the temperature when the reading on the voltmeter is 2.8 mV. Give your answer in degrees Celsius and in kelvin.

  (You may assume that the thermo-e.m.f produced is directly proportional to the temperature difference between the junctions.)

Specific heat capacity and specific latent heat

Specific heat capacity

When an object is heated, its temperature increases. The amount that it increases by ($\Delta\theta$) depends on:

» the energy supplied (ΔQ)
» the mass of the object (m)
» the material the object is made from

$$\Delta\theta \propto \frac{\Delta Q}{m}$$

which can be written:

$$\Delta Q = mc\Delta\theta$$

where c is the constant of proportionality. Its value depends on the material which is being heated. This is known as the **specific heat capacity** of the material.

Rearranging the equation gives:

$$c = \frac{\Delta Q}{m\Delta\theta}$$

The units of specific heat capacity are $J\,kg^{-1}\,K^{-1}$, although $J\,kg^{-1}\,°C^{-1}$ is often used. The units are numerically equal.

> **KEY TERMS**
>
> The **specific heat capacity** of a material is the energy required to raise the temperature of unit mass of the material by unit temperature.

WORKED EXAMPLE

An electric shower is designed to work from a 230 V mains supply. It heats the water as it passes through narrow tubes prior to the water passing through the shower head. Water enters the heater at 12°C and when the flow rate is 0.12 kg s⁻¹ it leaves at 28°C. Calculate the current in the heater, assuming that energy losses are negligible.
(specific heat capacity of water = 4200 J kg⁻¹ °C⁻¹)

Answer

power = $VI = mc\Delta\theta$

where m is the mass of water passing through the heater per second

$230I = 0.12 \times 4200 \times (28 - 12)$

$I = 35\,A$

3 A block of aluminium has a mass of 0.50 kg. It is heated, with a 36 W heater, for 3 minutes and its temperature increases from 12°C to 26°C. Calculate the specific heat capacity of aluminium.

4 500 g of copper rivets are placed in a polystyrene cup and are heated using a 40 W heater. The initial temperature of the rivets is 12°C. After heating for 5 minutes, the temperature rises to 70°C. Calculate the specific heat capacity of copper.

Specific latent heat

» You will have observed that when a beaker of water boils it remains at a constant temperature of 100°C, despite energy still being supplied.
» The energy supplied does not change the temperature of the substance.
» Instead, it is doing work in separating the molecules of water, changing it from liquid to vapour.
» This energy is called the **latent heat of vaporisation**.
» Similarly, when ice melts it remains at a constant temperature of 0°C and the supplied energy changes the solid to liquid.
» This energy is called the **latent heat of fusion**.

From the definitions:

$$L_f = \frac{\Delta Q}{\Delta m}$$

where L_f is the specific latent heat of fusion, ΔQ is the energy input and Δm is the mass of solid converted to liquid.

$$L_v = \frac{\Delta Q}{\Delta m}$$

where L_v is the specific latent heat of vaporisation, ΔQ is the energy input and Δm is the mass of liquid converted to vapour.

The units of both specific latent heat of fusion and of vaporisation are $J\,kg^{-1}$.

> **KEY TERMS**
>
> The **specific latent heat of fusion** is the energy required to change unit mass of solid to liquid without change in temperature.
>
> The **specific latent heat of vaporisation** is the energy required to change unit mass of liquid to vapour without change in temperature.

WORKED EXAMPLE

A 1.5 kW kettle contains 400 g of boiling water. Calculate the mass of water remaining if it is left switched on for a further 5 minutes. (specific latent heat of vaporisation of water = 2.26 MJ kg^{-1})

Answer

$$L_v = \frac{\Delta Q}{\Delta m}$$

Therefore:

$$\Delta m = \frac{\Delta Q}{L_v} = \frac{(1.5 \times 10^3) \times (5 \times 60)}{2.26 \times 10^6} = 0.199\,kg = 199\,g$$

mass remaining = 400 − 199 = 201 g

▶ NOW TEST YOURSELF

5   The apparatus shown in Figure 14.2 is used to measure the specific latent heat of fusion of ice.
The heater has an output power of 60 W.
The initial reading on the balance is 125.0 g.
The reading after 5 minutes is 179.5 g.
Calculate the specific latent heat of fusion of ice.

6   A 1.25 kW kettle containing 0.75 kg of water at 20°C is switched on. Calculate the mass of water left in the kettle after 8 minutes. (specific heat capacity of water = 4200 J kg^{-1} K^{-1}, specific latent heat of vaporisation of water = 2 260 000 J kg^{-1})

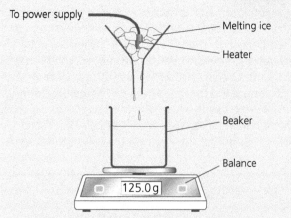

▲ **Figure 14.2 Apparatus to measure the specific latent heat of fusion of ice**

REVISION ACTIVITY

Look up, on the internet, the energy arriving at the Earth from the Sun per day. How much ice could this energy melt? Fortunately, most of this energy is reradiated into space. If there were a 0.1% decrease in the amount of energy reradiated away, what extra mass of ice could be melted in a year?

END OF CHAPTER CHECK

In this chapter, you have learnt to:

» understand that energy is transferred from a region of higher temperature to a region of lower temperature ☐

» understand that regions of equal temperature are in thermal equilibrium ☐

» understand that a physical property that varies with temperature may be used to measure temperature and state examples of such properties ☐

» understand that the scale of the thermodynamic temperature does not depend on the property of any particular substance ☐

» convert temperatures between kelvin and degrees Celsius ☐

» recall that $T/K = \theta/°C + 273.15$ ☐

» understand that the lowest possible temperature is zero kelvin on the thermodynamic temperature scale and that this is known as absolute zero ☐

» define and use specific heat capacity ☐

» recall and use the equation $c = \Delta Q/m\Delta\theta$ ☐

» define and use specific latent heat and distinguish between specific latent heat of fusion and specific latent heat of vaporisation ☐

» recall and use the equation $L = \Delta Q/\Delta m$ ☐

15 Ideal gases

The mole and the Avogadro constant

» You are already familiar with the idea of measuring mass in kilograms and thinking of mass in terms of the amount of matter in an object.
» The **mole** measures the amount of matter from a different perspective – the number of particles in an object.
» One **mole** was traditionally defined as the amount of substance that has the same number of particles as there are atoms in 12 g of carbon-12 isotope. (See p. 94 for information about isotopes.)
» The more modern definition states the precise number of particles in a mole. This number is the Avogadro constant (symbol N_A) and is equal to $6.02214076 \times 10^{23}$.
» The amount of matter is a base quantity and its unit the mole, consequently, is a base unit.
» The abbreviation for the mole is **mol**.

So:

» One mole of carbon-12 isotope contains 6.02×10^{23} carbon-12 atoms and has a mass of 12 g.
» One mole of helium-4 isotope contains 6.02×10^{23} helium-4 atoms and has a mass of 4 g.

Many gases are found not as single atoms but as diatomic molecules. For example, two hydrogen atoms form a H_2 molecule, so one mole of hydrogen contains 6.02×10^{23} hydrogen molecules (H_2) or 12.04×10^{23} atoms of hydrogen.

> **STUDY TIP**
>
> You are not expected to remember the Avogadro constant to 9 significant figures! Generally, we shall work to no more than 3 significant figures, making the Avogadro constant 6.02×10^{23}.

WORKED EXAMPLE

Calculate the number of atoms in, and the mass of, the following:

a 1 mol of ozone (O_3)
b 3 mol of water (H_2O)
   (relative atomic mass of oxygen = 16, relative atomic mass of hydrogen = 1)

Answer

a 1 mol of ozone = 6.02×10^{23} molecules
                  = $3 \times 6.02 \times 10^{23}$ atoms
                  = 18.06×10^{23} atoms $\approx 18.1 \times 10^{23}$ atoms
   mass of ozone in 1 mol = $3 \times 16 = 48$ g
b Each molecule of water contains 3 atoms (2 hydrogens, 1 oxygen).
   number of atoms in 1 mol of water = $3 \times 6.02 \times 10^{23} = 18.06 \times 10^{23}$ atoms
   number of atoms in 3 mol = $3 \times 18.06 \times 10^{23} \approx 5.42 \times 10^{24}$ atoms
   1 mol of water has mass = $(2 \times 1) + (1 \times 16) = 18$ g
   Therefore, the mass of 3 moles = $3 \times 18 = 54$ g.

▶ NOW TEST YOURSELF   TESTED ☐

1 The mass of 1 mol of hydrogen gas is 2 g. Calculate the mass of 1 hydrogen atom.
2 Calculate the number of atoms of each element in 1 mol of the chemical Na_2CO_3.

Equation of state

The ideal gas equation

Experimental work shows that a fixed mass of any gas, at temperatures well above the temperature at which it condenses to form a liquid, and at a wide range of pressures, follows the following relationships:

» at constant temperature, $p \propto \dfrac{1}{V}$
» at constant pressure, $V \propto T$
» at constant volume, $p \propto T$

Combining the three proportionalities, it follows that:

$$pV \propto T$$

where p is the pressure, V is the volume and T is the temperature measured on the kelvin scale. The kelvin scale of temperature is discussed further on p. 128.

These three relationships can be combined to form a single equation:

$$\frac{pV}{T} = \text{constant}$$

The equation can be written as:

$$pV = nRT$$

where n is the number of moles of gas and R is the molar gas constant. The molar gas constant has the same value for all gases, $8.31\,\text{J K}^{-1}$. This equation is known as the equation of state for an ideal gas.

> An ideal gas is defined as a gas that obeys the equation of state, $pV = nRT$, at all temperatures, pressures and volumes.

Real gases, such as hydrogen, helium and oxygen, follow the equation at room temperature and pressure. However, if the temperature is greatly decreased or the pressure is very high, they no longer behave in this way.

WORKED EXAMPLE

1 A syringe of volume $25\,\text{cm}^3$ holds hydrogen at a pressure of $1.02 \times 10^5\,\text{Pa}$ and temperature $280\,\text{K}$. The volume of the gas is reduced to $10\,\text{cm}^3$ and the temperature increases by $5\,\text{K}$. Calculate the new pressure of the gas.

Answer

Can be rewritten as $\dfrac{p_1 V_1}{T_1} = \dfrac{p_2 V_2}{T_2}$

Substitute in the values:

$$\frac{(1.02 \times 10^5) \times 25}{280} = \frac{p_2 \times 10}{285}$$

Thus:

$$p_2 = 2.6 \times 10^5\,\text{Pa}$$

2 Calculate the volume occupied by $48\,\text{mg}$ of oxygen at $20°C$ and a pressure of $1.0 \times 10^5\,\text{Pa}$. (relative atomic mass of oxygen = 16)

Answer

temperature $= 273 + 20 = 293\,\text{K}$

oxygen forms diatomic O_2 molecules, so the mass of 1 mol of oxygen $= 32\,\text{g}$

number of moles in $48\,\text{mg} = \dfrac{48 \times 10^{-3}}{32} = 1.5 \times 10^{-3}\,\text{mol}$

Using $pV = nRT$:

$$V = \frac{nRT}{p} = \frac{(1.5 \times 10^{-3}) \times 8.3 \times 293}{1.0 \times 10^5} = 3.7 \times 10^{-5}\,\text{m}^3$$

3 An ideal gas is held in a syringe of volume 200 cm^3 at a pressure of
   4.5×10^5 Pa. The gas is allowed to expand until it reaches a pressure of
   1.02×10^5 Pa. As the gas expands, its temperature falls from 300 K to 280 K.
   Calculate the volume the gas will now occupy.
4 The pressure in a helium-filled party balloon of volume of 0.060 m^3 is
   0.12 MPa at a temperature of 22°C. Calculate:
   a the number of moles of helium
   b the mass of helium in the balloon
      (the mass of a helium atom = 4 u)

The Boltzmann constant   REVISED ☐

An alternative way of writing the equation of state of an ideal gas is:

$pV = NkT$

where N is the number of molecules and k is the Boltzmann constant.

Compare the two equations for 1 mol of an ideal gas:

$pV = RT$

$pV = N_A kT$

Therefore, $N_A k = R$ and therefore $k = R/N_A$.

> **NOW TEST YOURSELF**   TESTED ☐

5 Given that the value of the molar gas constant R is 8.31 J K^{-1}, calculate the
   value of the Boltzmann constant, giving its units.

Kinetic theory of gases

A model of a gas   REVISED ☐

- » From earlier studies at IGCSE or a similar course, you should be able to picture a
  gas as small independent particles moving freely at random. This is described as
  a 'model' of a gas.
- » An ideal gas is modelled as consisting of many molecules that have random
  velocities – this means they move in random directions with random speeds.
- » The molecules themselves can be modelled as small unbreakable spheres that are
  spaced well apart and only interact with one another when they collide.
- » If the gas is enclosed in a container, a pressure is produced on the container.
- » This is due to the molecules colliding elastically with the container walls and
  the change in momentum of the particles producing many little impulses on the
  walls.
- » There are a large number of impulses per unit time which produce an average
  force on an area of the wall.
- » Pressure is force per unit area.

Relationship between molecular speed and pressure exerted

To show the relationship between the speed of the molecules in a gas and the pressure the gas exerts, the following assumptions are made:

» The forces between the molecules are negligible (except during collisions).
» The volume of the molecules is negligible compared with the total volume occupied by the gas.
» All collisions between the molecules, and between the molecules and the container walls, are perfectly elastic.
» The time spent in colliding is negligible compared with the time between collisions.
» There are many identical molecules that all move at random.

> **STUDY TIP**
>
> These assumptions effectively describe an ideal gas.

Consider a gas molecule of mass m in a cubic box of side L. The molecule is travelling at speed c parallel to the base of the box (Figure 15.1).

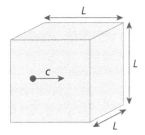

▲ Figure 15.1

When the molecule collides with the right-hand wall it will rebound with velocity $-c$.

change in momentum $= -2mc$

The molecule travels a distance of $2L$ before colliding with that wall again, so the time elapsed is $2L/c$.

rate of change of momentum = force applied by the molecule on this wall

$$= \frac{2mc}{2L/c} = \frac{mc^2}{L}$$

The area of the wall is L^2, so:

$$\text{pressure} = \frac{\text{force}}{\text{area}} = \frac{mc^2}{L^3}$$

The molecule being considered is moving perpendicular to the two faces with which it collides. In practice, a typical molecule moves randomly and collides with all six faces. Thus, the total area involved is three times that which has been considered, so:

$$\text{pressure} = \frac{mc^2}{3L^3}$$

The total number of molecules in the box is N, each with a different speed c contributing to the overall pressure. The average of the velocities squared is called the mean-square speed, $<c^2>$.

So:

$$\text{pressure} = \frac{1}{3} \frac{Nm<c^2>}{L^3}$$

$L^3 = V$, the volume of the box.

$$p = \frac{1}{3} \frac{Nm<c^2>}{V}$$

It is sometimes useful to write this equation as:

$$pV = \frac{1}{3}Nm<c^2>$$

The root-mean-square speed

The root-mean-square speed is the square root of the mean (or average) of the square of the velocity.

$$c_{r.m.s.} = \sqrt{<c^2>}$$

»» Velocity is a vector.
»» The molecules are moving in all directions and so the sum of all the velocities is zero.
»» The average velocity of many, many molecules is zero.
»» We square the velocities as this gives positive numbers only, hence the positive square root of this number is greater than zero.

WORKED EXAMPLE

At room temperature and pressure (293 K and 1.0×10^5 Pa), 1 mol of any gas occupies a volume of 24 dm^3. Calculate the root-mean-square speed of the following at this temperature:

a helium atoms (atomic mass = 4 u)
b oxygen molecules (atomic mass = 16 u, mass of O_2 = 32 u)

Answer

a Nm = total mass of 1 mol of helium = 4×10^{-3} kg

$$p = \frac{1}{3}\frac{Nm<c^2>}{V}$$

$$<c^2> = \frac{3pV}{Nm} = \frac{3 \times (1.0 \times 10^5) \times (24 \times 10^{-3})}{4 \times 10^{-3}} = 18 \times 10^5 \, m^2 s^{-2}$$

$$\sqrt{<c^2>} = 1342 \, m s^{-1} = 1300 \, m s^{-1}$$

b Nm = total mass of 1 mol of O_2 molecules = 3.2×10^{-2} kg

$$<c^2> = \frac{3pV}{Nm} = \frac{3 \times (1.0 \times 10^5) \times (24 \times 10^{-3})}{3.2 \times 10^{-3}} = 2.25 \times 10^5 \, m^2 s^{-2}$$

$$\sqrt{<c^2>} = 470 \, m s^{-1}$$

▶ NOW TEST YOURSELF

TESTED ☐

6 Five molecules have speeds of $350 \, m s^{-1}$, $361 \, m s^{-1}$, $425 \, m s^{-1}$, $284 \, m s^{-1}$ and $620 \, m s^{-1}$. Calculate:
   a their mean-square speed
   b their root-mean-square speed
7 Calculate the root-mean-square speed of nitrogen molecules at 0°C. (mass of nitrogen molecule = 4.6×10^{-26} kg)

Kinetic energy of a molecule

Temperature and molecular kinetic energy

If we compare the ideal gas equation ($pV = nRT$) and the equation $pV = \frac{1}{3}Nm<c^2>$, we can see that:

$$nRT = \frac{1}{3}Nm<c^2>$$

Check your answers at www.hoddereducation.com/cambridgeextras

For one mole:

$$\frac{RT}{N_A} = \frac{1}{3}m{<}c^2{>}$$

$$kT = \frac{1}{3}m{<}c^2{>}$$

where k is the **Boltzmann constant**, and hence:

$3kT = m{<}c^2{>}$ and

$$\frac{3}{2}kT = \frac{1}{2}m{<}c^2{>}$$

where $\frac{1}{2}m{<}c^2{>}$ is equal to the average (translational) kinetic energy of the molecules. Hence, the temperature is proportional to the average (translational) kinetic energy of the particles in a gas.

> ### KEY TERMS
>
> The **Boltzmann constant** $k = \dfrac{R}{N_A}$, and has the value $1.38 \times 10^{-23}\,\text{J K}^{-1}$.

▶ NOW TEST YOURSELF   TESTED ☐

8 Calculate the average translational kinetic energy of air molecules when the air temperature is 20°C.

▶ REVISION ACTIVITIES

The relationship between temperature and kinetic energy only works precisely for an ideal gas. For a monatomic gas, it works well, but less well for diatomic or triatomic gases. Suggest explanations for these facts.

'Must learn' equations:

$pV = nRT$

$pV = \frac{1}{3}Nm{<}c^2{>}$

$\frac{3}{2}kT = \frac{1}{2}m{<}c^2{>}$

$\frac{R}{N_A} = \frac{1}{3}Nm{<}c^2{>}$

▶ END OF CHAPTER CHECK

In this chapter, you have learnt:
- ▸ that the unit for the amount of substance is the mole (mol) ☐
- ▸ that the mol is the unit of a base quantity and is, itself, a base unit ☐
- ▸ to understand that a gas obeying $pV \propto T$, where T is the thermodynamic temperature, is known as an ideal gas ☐
- ▸ to recall and use the equation of state as $pV = nRT$, where n is the amount of substance, and as $pV = NkT$, where N is the number of molecules ☐

- ▸ to recall that the Boltzmann constant k is given by $k = R/N_A$ ☐
- ▸ to state the basic assumptions of the kinetic theory of gases ☐
- ▸ to explain how molecular movement causes the pressure exerted by a gas ☐
- ▸ to derive and use the relationship $pV = \frac{1}{3}Nm{<}c^2{>}$, where ${<}c^2{>}$ is the mean-square speed ☐
- ▸ to understand that the root-mean-square speed $c_{\text{r.m.s.}} = \sqrt{{<}c^2{>}}$ ☐
- ▸ to compare $pV = \frac{1}{3}Nm{<}c^2{>}$ with $pV = NkT$ to deduce the average translational kinetic energy of a molecule is $\frac{3}{2}kT$ ☐

Internal energy and the first law of thermodynamics

Internal energy

REVISED ☐

» In the previous sections, we have seen that the particles in an object have a mixture of kinetic energy and potential energy.
» Kinetic energy determines the temperature of the object, and potential energy determines the state of the object.
» Not all particles have the same kinetic and potential energies – they are randomly distributed.
» The **internal energy** of an object is the sum of the random kinetic and potential energies of all the particles in the object.

> **KEY TERMS**
> **Internal energy** is the sum of a random distribution of the kinetic and potential energies associated with the molecules of a system.

The first law of thermodynamics

REVISED ☐

There are two ways of increasing the total internal energy of an object:

» heating the object
» doing work on the object

This leads to the first law of thermodynamics, which can be expressed by the equation:

increase in internal energy (ΔU) = the energy supplied to the system by heating (q) + the work done on the system (W)

$$\Delta U = q + W$$

(The 'energy supplied to the system by heating' is sometimes shortened to the 'heating of the system'.)

To demonstrate a use of the first law, consider an ideal gas contained in a cylinder by a frictionless piston. The initial volume of gas is V (Figure 16.1).

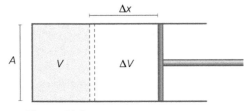

▲ Figure 16.1

The gas is heated so that its volume increases by an amount ΔV against a constant atmospheric pressure. The gas expands, so work is done against atmospheric pressure:

$$W = F\Delta x$$

where F is the force on the piston.

force on the piston = pressure of the gas × area of cross-section of the piston = pA

So:

$$W = pA\Delta x = p\Delta V$$

Applying the first law of thermodynamics:

$\Delta U = q + W$

$\Delta U = q - p\Delta V$

» The minus sign comes in because the work is done by the gas on the atmosphere, rather than work being done on the gas.
» Notice that the change in internal energy is less than the thermal energy input q, because some of the energy is used to do work in expanding the gas.
» It should also be understood that when the internal energy of a system increases, the temperature increases and when the internal energy decreases, the temperature decreases.

Table 16.1 explains when the quantities ΔU, q and W should be considered positive and negative.

▼ Table 16.1

Quantity	Positive	Negative
ΔU	The internal energy increases	The internal energy decreases
q	Energy is transferred to the system from the surroundings by heating	Energy is transferred from the system to the surroundings by heating
W	Work is done on the system	Work is done by the system

WORKED EXAMPLE

$0.14\,m^3$ of helium at a temperature of 20°C is contained in a cylinder by a frictionless piston. The atmospheric pressure is a steady $1.02 \times 10^5\,Pa$.

The helium is heated until the temperature of the gas is 77°C. The piston moves so that the pressure on the gas remains at atmospheric pressure.

a Calculate the change in volume of the helium at 77°C.

b Calculate the work done by the gas on the atmosphere.

Answer

a Consider helium to be an ideal gas. Therefore:

$$\frac{(p_1V_1)}{T_1} = \frac{(p_2V_2)}{T_2} \rightarrow \frac{V_1}{T_1} = \frac{V_2}{T_2}$$

p remains constant throughout the change and therefore p_1 and p_2 cancel.

Substitute in the values:

$0.14/(20 + 273) = V_2/(77 + 273)$

$V_2 = (0.14 \times 350)/293 = 0.167\,m^3$

$\Delta V = V_2 - V_1 = 0.167 - 0.14 = 0.027\,m^3$

b $\Delta W = p\Delta V = 1.02 \times 10^5 \times (0.167 - 0.140) = 2800\,J$

NOW TEST YOURSELF

TESTED ☐

1 An isothermal change is a change in which there is no change in temperature. Explain why there can be a change in internal energy of a system in an isothermal change of a real gas.

REVISION ACTIVITY

'Must learn' equations:

$\Delta U = q + W$

$W = p\Delta V$

END OF CHAPTER CHECK

In this chapter, you have learnt to:
» understand that the internal energy is determined by the state of the system, and it can be expressed as the sum of the random distribution of kinetic and potential energies associated with the molecules of the system ☐
» relate a rise in temperature of an object to an increase in its internal energy ☐

» recall and use $W = p\Delta V$ for the work done when the volume of a gas changes at constant pressure ☐
» understand the difference between the work done by the gas and the work done on the gas ☐
» recall and use the first law of thermodynamics, $\Delta U = q + W$, expressed in terms of the increase in internal energy, the heating of the system and the work done on the system ☐

17 Oscillations

Simple harmonic oscillations

Terminology

Consider a ruler clamped to a bench, pulled downwards and released so that it vibrates; a pendulum swinging backwards and forwards; a mass on the end of a spring bouncing up and down. These are all examples of oscillating systems (Figure 17.1).

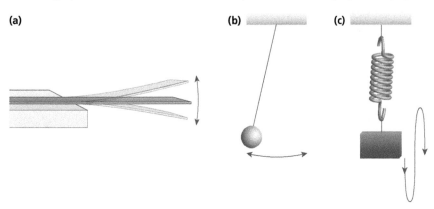

(a)   **(b)**   **(c)**

▲ Figure 17.1 Oscillating systems

One complete oscillation is when an object moves:

» from its equilibrium position to its maximum displacement in one direction
» back through the equilibrium position to the maximum displacement in the opposite direction
» and back once more to the equilibrium position

This is shown in Figure 17.1(c).

» The **period**, T, is the time taken for one complete oscillation of an object.
» The **frequency**, f, is the number of oscillations per unit time.
» The **displacement**, x, is the vector distance from the equilibrium position at an instant.
» The **amplitude**, x_0, is equal to the magnitude of the maximum displacement of an object from its mean position.
» The **angular frequency**, ω, is equal to $2\pi f$.
» The **phase difference** is the fraction of a cycle between two oscillating objects, expressed in either degrees or radians. (See p. 57.)

It is worth remembering the following relationships, which you may recognise from the work on circular motion:

$$F = \frac{1}{T} \qquad\qquad \omega = 2\pi f \qquad\qquad \omega = \frac{2\pi}{T}$$

Simple harmonic oscillations

In the examples above, the objects vibrate in a particular way known as **simple harmonic motion (s.h.m.)**. There are many other types of oscillations. For instance, a conducting sphere will oscillate between two charged conducting plates – but not with simple harmonic motion.

The conditions required for simple harmonic motion are:

» the magnitude of the acceleration is proportional to the displacement from a fixed point
» the direction of the acceleration is always in the opposite direction to the displacement
» this means that the acceleration is always directed towards the fixed point

Simple harmonic motion can be investigated using a position sensor connected to a datalogger (Figure 17.2).

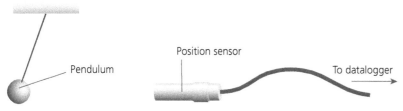

Pendulum   Position sensor   To datalogger

▲ Figure 17.2

The displacement against time graph can be deduced from the trace on the datalogger (Figure 17.3).

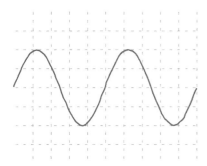

▲ Figure 17.3

As with any displacement–time graph, the velocity is equal to the gradient of the graph; the acceleration is equal to the gradient of the velocity–time graph.

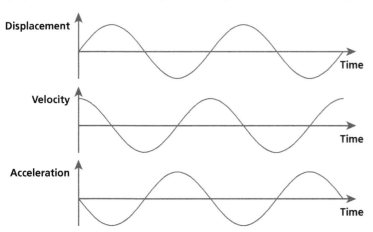

Displacement — Time
Velocity — Time
Acceleration — Time

▲ Figure 17.4

Table 17.1 describes the displacement, the velocity and the acceleration at different points during an oscillation, with reference to Figure 17.4.

» A more detailed analysis of the graphs shows that the velocity–time graph is obtained by plotting a graph of the gradient of the distance–time graph against time. The equation for this graph is $v = v_0 \cos \omega t$
» The acceleration–time graph is obtained by plotting a graph of the gradient of the velocity–time graph against time.
» Remember that the velocity is obtained from the gradient of the distance–time graph and acceleration is from the gradient of the velocity–time graph. Refer back to p. 19.

> **STUDY TIP**
>
> In the conducting sphere example, when the charged sphere bounces back and forth between two parallel charged conductors, the magnitude of the electric force is constant as it moves between the plates. The first condition (acceleration ∝ displacement) is not followed. Therefore, this cannot be simple harmonic motion.

> **STUDY TIP**
>
> The displacement–time graph can be started at any point on the cycle. Here, the equilibrium position is chosen as the starting point. Other books might choose maximum displacement, in which case the displacement curve would be a cosine curve, the velocity curve would be a minus sine curve and the acceleration would be a minus cosine curve.

Point in cycle	Displacement	Velocity	Acceleration
$t = 0$	Zero	Maximum in one direction	Zero
¼ cycle on from $t = 0$	Maximum in one direction	Zero	Maximum in the opposite direction from the displacement
½ cycle on from $t = 0$	Zero	Maximum in the opposite direction from before	Zero
¾ cycle on from $t = 0$	Maximum in the opposite direction to before	Zero	Maximum in the opposite direction from the displacement
1 cycle on from $t = 0$	Zero	Maximum in the original direction	Zero

Equations for simple harmonic motion

REVISED ☐

If you look at the graphs of simple harmonic motion (s.h.m.), you will see that they are of the form of sine (or cosine) graphs. The conditions for s.h.m. give the following proportionality:

$$a \propto -x$$

where a is the acceleration and x is the displacement.

The minus sign comes in because the acceleration is in the opposite direction from the displacement.

This leads to the equation:

$$a = -\omega^2 x$$

where ω is the angular frequency.

This equation describes simple harmonic motion. The graphs in Figure 17.4 are 'solutions' to this equation. If you look at those graphs, you will see that they have a sine (or cosine) shape. The precise equations that they represent are:

» displacement: $x = x_0 \sin \omega t$
» velocity: $v = x_0 \omega \cos \omega t$
» acceleration: $a = -x_0 \omega^2 \sin \omega t$

where x_0 is the amplitude of the oscillation.

Look at the equations for displacement and acceleration. Can you see that they fit in with the equation $a = -\omega^2 x$?

The velocity of the vibrating object at any point in the oscillation can be calculated using the formula:

$$v = \pm \omega \sqrt{x_0^2 - x^2}$$

It follows that when $x = 0$ (i.e. the displacement is zero) the velocity is a maximum and:

$$v_0 = \pm \omega x_0$$

STUDY TIP

$a = -x_0 \omega^2 \sin \omega t$ and $x = x_0 \sin \omega t$

Dividing the first equation by the second gives:

$$\frac{a}{x} = \frac{-x_0 \omega^2 \sin \omega t}{x_0 \sin \omega t}$$

Cancelling the x_0 and the $\sin \omega t$ top and bottom gives:

$$\frac{a}{x} = -\omega^2$$

which gives:

$$a = -\omega^2 x$$

WORKED EXAMPLE

A mass on the end of a spring oscillates with a period of 1.6 s and an amplitude of 2.4 cm. Calculate:

a   the angular frequency of the oscillation

b   the maximum speed of the mass

c   the maximum acceleration

d   the speed of the mass when its displacement from the equilibrium position is 0.60 cm

Answer

a $f = \dfrac{1}{T} = \dfrac{1}{1.6}\,\text{Hz}$

   $\omega = 2\pi f = 2\pi \times \dfrac{1}{1.6} = 3.9\,\text{rad s}^{-1}$

b $v_{max} = \omega x_0 = 3.9 \times 2.4 = 9.4\,\text{cm s}^{-1}$

c $a = -\omega^2 x$

   $a_{max} = \omega^2 x_0 = 3.9^2 \times 2.4 = 37\,\text{cm s}^{-2}$

d $v = \pm\,\omega\sqrt{x_0^2 - x^2} = 3.9 \times \sqrt{2.4^2 - 0.6^2} = 9.1\,\text{cm s}^{-1}$

NOW TEST YOURSELF

TESTED ☐

1   A stone of mass 0.80 kg attached to the bottom of a vertical spring oscillates with a time period of 1.8 s and an amplitude of 4.4 cm. Calculate:

   a   the frequency of the stone

   b   the angular frequency of the stone

   c   the maximum acceleration of the stone

2   The period of a simple pendulum is given by the formula: $T = 2\pi\sqrt{l/g}$.

   a   Calculate the period of a pendulum of length 25 cm.

   b   Calculate the acceleration of the bob when the displacement is 3.0 cm.

3   A particle vibrates with simple harmonic motion of amplitude 5.0 cm and frequency 0.75 Hz. Calculate the maximum speed of the particle and its speed when it is 2.5 cm from the central position.

---

s.h.m. and circular motion

REVISED ☐

The introduction of ω should have reminded you of circular motion. The description of the following experiment shows the relationship between circular motion and simple harmonic motion (Figure 17.5).

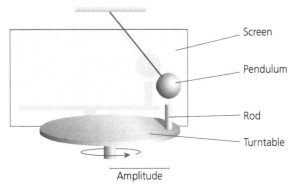

Screen

Pendulum

Rod

Turntable

Amplitude

▲ **Figure 17.5**

»   A rod is set up on a turntable, which rotates.

»   A pendulum is set swinging with an amplitude equal to the radius of the rotation of the rod.

»   The speed of rotation of the turntable is adjusted until the time for one revolution of the turntable is exactly equal to the period of the pendulum.

- The whole apparatus is illuminated from the front so that a shadow image is formed on a screen.
- It is observed that the shadow of the pendulum bob moves exactly as the shadow of the rod.
- This shows that the swinging of the pendulum is the same as the projection of the rod on the diameter of the circle about which it rotates (Figure 17.6).

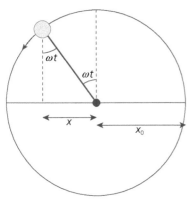

▲ Figure 17.6

You should now understand the close mathematical relationship between circular motion and simple harmonic motion.

Energy in simple harmonic motion

Kinetic energy and potential energy

During simple harmonic motion, energy is transferred continuously between kinetic and potential energy:

- In the case of a pendulum, the transfer is between kinetic and gravitational potential energy.
- In the case of a mass tethered between two horizontal springs, the transfer is between kinetic and strain potential energy.

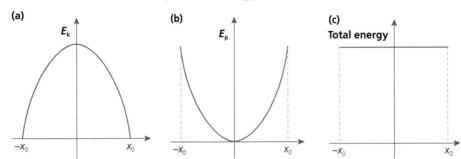

▲ Figure 17.7 (a) The variation of kinetic energy with displacement, (b) the variation of potential energy with displacement, (c) the total energy with displacement

- The speed of the particle is at a maximum when the displacement is zero so that the kinetic energy is maximum at this point and the potential energy is zero (Figure 17.7a,b).
- At maximum displacement, the speed, and hence the kinetic energy, is zero and the potential energy is maximum (Figure 17.7a,b).
- The important point is that in any perfect simple harmonic oscillator, the total energy is constant. This means that the sum of the kinetic and potential energies remains constant throughout each oscillation (Figure 17.7c).

The equations that link the kinetic energy and the potential energy to the displacement are:

- kinetic energy: $E_k = \frac{1}{2}m\omega^2(x_0^2 - x^2)$
- potential energy: $E_p = \frac{1}{2}m\omega^2x^2$
- total energy at any point in the oscillation: $E_k + E_p \Rightarrow E = \frac{1}{2}m\omega^2x_0^2$

Check your answers at **www.hoddereducation.com/cambridgeextras**

WORKED EXAMPLE

A clock pendulum has a period of 2.0 s and a mass of 600 g. The amplitude of the oscillation is 5.2 cm. Calculate the maximum kinetic energy of the pendulum and, hence, its speed when it is travelling through the centre point.

Answer

$E_k = \frac{1}{2}m\omega^2(x_0^2 - x^2)$; for maximum speed the displacement = 0

$T = 2.0$, therefore $\omega = \dfrac{2\pi}{2} = \pi$

$E_k = \frac{1}{2}m\omega^2 x_0^2 = 0.5 \times 0.600 \times \pi^2 \times (5.2 \times 10^{-2})^2$

$E_k = 8.0 \times 10^{-3}$ J

$E_k = \frac{1}{2}mv^2$

$v = \sqrt{\dfrac{2E_k}{m}} = \sqrt{\dfrac{2 \times (8.0 \times 10^{-3})}{0.6}} = 0.16\,\text{m s}^{-1}$

NOW TEST YOURSELF

TESTED ☐

4 A pendulum has a length of 5.0 m and an amplitude of 12 cm. The bob has a mass of 0.50 kg. Calculate:

  a the maximum speed of the pendulum bob

  b the maximum restoring force on the bob

  c the maximum kinetic energy of the bob

  d the total energy of the system

Damped and forced oscillations, resonance

Damping

REVISED ☐

» Up to this point, we have only looked at perfect simple harmonic motion, where the total energy is constant and no energy is lost to the surroundings.

» In this situation, where the only force acting on the oscillator is the restoring force, the system is said to be in **free oscillation**.

» In real systems, some energy is lost to the surroundings for a variety of reasons, including due to friction and/or air resistance.

» This always acts in the opposite direction to the restoring force.

» The result is that the amplitude of the oscillations gradually decreases. This is called **damping** (Figure 17.8).

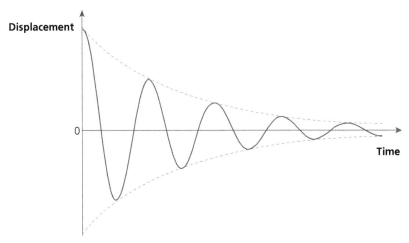

▲ Figure 17.8 A lightly damped oscillation

» The decay of the oscillation follows the exponential decay (see p. 161).

» The period, however, remains constant until the oscillation dies away completely.

» Figure 17.8 shows **light damping** – the oscillation gradually fades away.

» If the damping is increased, we eventually reach a situation where no complete oscillations occur and the displacement falls to zero. When this occurs in the minimum time, the damping is said to be **critical** (see Figure 17.9).

» More damping than this is described as **heavy damping** and the displacement only slowly returns to zero (see Figure 17.9).

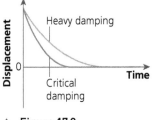

▲ Figure 17.9

Examples of damped oscillations

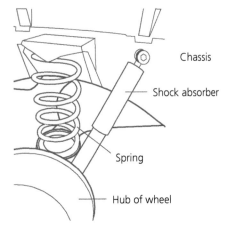

▲ **Figure 17.10 The suspension on a car relies on critically damped harmonic motion**

A car suspension (Figure 17.10) operates in a critical damping mode in order to bring the displacement back to zero in the shortest possible time without oscillations. A heavily damped suspension leads to a hard ride, with energy given to the car by bumps not being absorbed as efficiently.

Forced oscillations

REVISED ▢

» In Chapter 8, you met the idea of stationary waves formed on a string (pp. 67–68). This is an example of a forced oscillation.

» An extra periodic force is applied to the system. This periodic force continuously feeds energy into the system to keep the vibration going.

» You will have observed how the amplitude of the vibrations of the waves on a string changes as the frequency of the vibrator is changed:
  » a small amplitude at very low frequencies
  » gradually increasing to a maximum as the frequency is increased
  » then reducing again as the frequency is increased further (Figure 17.11)

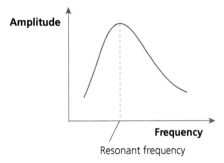

▲ **Figure 17.11 The amplitude of a forced oscillation at different frequencies**

» This is an example of **resonance**.

» When the driving frequency is the same as the **natural frequency** of oscillation of the string, then it gives the string a little kick at the right time each cycle and the amplitude builds up.

> **KEY TERMS**
>
> The **natural frequency** of a vibration is the frequency at which an object will vibrate when allowed to do so freely.

Resonance can be demonstrated using Barton's pendulums (Figure 17.12).

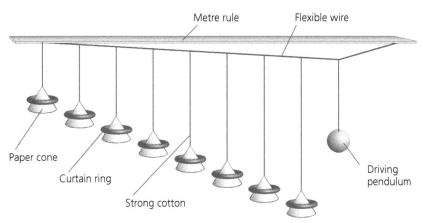

▲ **Figure 17.12 Barton's pendulums**

>> The driving pendulum causes the paper-cone pendulums to vibrate.
>> Only the pendulums of a similar length to the driving pendulum show any significant oscillation.
>> All the pendulums vibrate with the same frequency, which is the frequency of the driving pendulum (not their own natural frequencies).
>> This is a general rule for all forced oscillations.

NOW TEST YOURSELF

TESTED ☐

5 Explain the difference between critical damping and heavy damping.
6 Explain what is meant by the term resonance.

END OF CHAPTER CHECK

In this chapter, you have learnt to:
>> understand and use the terms displacement, amplitude, period, frequency and phase difference in the context of oscillations ☐
>> express the period in terms of frequency ☐
>> express the frequency in terms of angular frequency ☐
>> understand that simple harmonic motion occurs when acceleration is proportional to displacement from a fixed point and in the opposite direction ☐
>> use the formula $a = -\omega^2 x$ ☐
>> recall and use $x = x_0 \sin \omega t$ as a solution to the above equation ☐
>> use the equations $v = v_0 \cos \omega t$ and $v = \pm \omega \sqrt{x_0^2 - x^2}$ ☐

>> analyse and interpret graphical representations of the variations of displacement, velocity and acceleration for simple harmonic motion ☐
>> describe the interchange between kinetic and potential energy of a system undergoing simple harmonic motion ☐
>> understand that a resistive force acting on an oscillating system causes damping ☐
>> understand and use the terms light, critical and heavy damping and sketch displacement–time graphs illustrating these types of damping ☐
>> understand that resonance involves a maximum amplitude of oscillations and that this occurs when an oscillating system is forced to oscillate at its natural frequency ☐

18 Electric fields

Concept of an electric field

An electric field is a region in which charged objects experience a force.

> The **electric field strength** is defined as the force per unit positive charge on a stationary point charge.

$$\text{electric field strength} = \frac{\text{force}}{\text{charge}}$$

which can be written:

$$E = \frac{F}{q}$$

The unit of electric field strength is newtons per coulomb (NC^{-1}).

It is often useful to calculate the force on a charged particle in an electric field. In which case, the equation defining electric field can be rearranged to:

$$F = qE$$

Field lines

REVISED ▢

The shape of a field can be represented by drawing lines of force (Figure 18.1). In an electric field, the lines represent the direction of the force on a small positive test charge. When drawing an electric field:

» the direction of electric field lines is away from positive charges and towards negative charges
» the closer the field lines, the stronger the field strength
» the field lines never touch nor cross

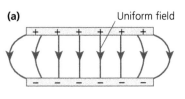

(a) Uniform field

Two oppositely charged parallel metal plates

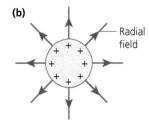

(b) Radial field

Positively charged sphere

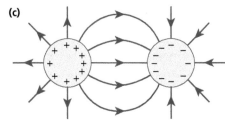

(c)

Two oppositely charged spheres

▲ **Figure 18.1 Shapes of electric fields**

Uniform electric fields

You can see from Figure 18.1(a) that once we move in from the edges of the plates, the field between two parallel plates is uniform. This means that wherever a charged particle is placed between those plates, it experiences the same magnitude force, in the same direction.

Calculating forces on charges

REVISED ▢

The electric field strength between the plates is given by the formula:

$$E = \frac{\Delta V}{\Delta d}$$

Check your answers at **www.hoddereducation.com/cambridgeextras**

where ΔV is the potential difference and Δd is the distance between the plates. Note that this means that an alternative way of expressing the unit of electric field strength ($N C^{-1}$) is volts per metre ($V m^{-1}$).

More generally, a uniform field is expressed in terms of the change in potential per unit charge.

WORKED EXAMPLE

A piece of dust carries a charge of -4.8×10^{-18} C, and lies at rest between two parallel plates separated by a distance of 1.5 cm. Calculate the force on the charge when a potential difference of 4500 V is applied across the plates.

Answer

$$E = \frac{V}{d} = \frac{4500}{1.5 \times 10^{-2}} = 300\,000\,V m^{-1}$$
$$E = \frac{F}{q}$$
$$F = E \times q$$
$$= 300\,000 \times (-4.8) \times 10^{-18}$$
$$= 1.44 \times 10^{-12}\,N$$

NOW TEST YOURSELF   TESTED ☐

1 Calculate the force on an electron when it is in an electric field of field strength $14\,kN C^{-1}$.
   (charge on an electron $e = 1.6 \times 10^{-19}$ C)
2 There is a potential difference of 5.0 kV across two parallel plates that are 2.0 cm apart. Calculate the electric field strength between the plates.
3 State the effect on the electric field strength in question 2 of:
   a using plates of two times the area of the original plates
   b moving one of the plates so that their separation is halved

Effect of electric fields on the motion of charged particles   REVISED ☐

» A charged particle in an electric field experiences a force and therefore tends to accelerate.
» If the particle is stationary or if the field is parallel to the motion of the particle, the magnitude of the velocity will change.
» An example is when electrons are accelerated from the cathode towards the anode in a cathode-ray tube (Figure 18.2).

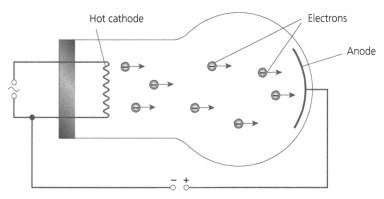

▲ **Figure 18.2 Principle of a cathode-ray tube**

- » If the field is at right angles to the velocity of the charged particles, the direction of the motion of the particles will be changed.
- » The path described by the charged particles will be parabolic (Figure 18.3) – the same shape as a projectile in a uniform gravitational field.
- » The component of the velocity perpendicular to the field is unchanged; the component of the velocity parallel to the field increases uniformly.

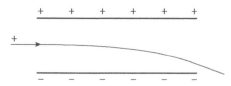

▲ Figure 18.3 The path of a proton as it passes through a uniform electric field

> **NOW TEST YOURSELF**   TESTED ☐

4 An electron is between two parallel metal plates which are 5.0 cm apart. A potential difference of 800 V is connected across the plates.
   a Calculate the electric field strength between the plates.
   b Calculate the acceleration of the electron.
   (charge on an electron = –1.6 × 10^{-19} C, mass of an electron = 9.1 × 10^{-31} kg)

Electric force between point charges

Coulomb's law   REVISED ☐

You will have met the idea that unlike charges attract and like charges repel in earlier courses. You may have deduced that the sizes of the forces between charges depend on:

- » the magnitude of the charges on the two objects
- » the distance between the two objects

The mathematical relationships are:

- » $F \propto Q_1 Q_2$, where Q_1 and Q_2 are the charges on the two objects.
- » $F \propto 1/r^2$ where r is the distance between the centres of the two charged objects

Combining the two relationships gives:

$$F \propto \frac{Q_1 Q_2}{r^2}$$

The constant of proportionality is $1/4\pi\varepsilon_0$, where ε_0 is known as the permittivity of free space and has a value 8.85 × 10^{-12} C^2 N^{-1} m^{-2}. This unit is often shortened to farads per metre (F m^{-1}). (We will meet the farad in the section on capacitors (see p. 156).)
So this gives:

$$F = \frac{Q_1 Q_2}{4\pi\varepsilon_0 r^2}$$

STUDY TIP

This relationship only really applies when point charges are considered. However, the charge on a spherical conductor can be considered to act at the centre of the conductor, provided the distance between the two objects is considerably greater than their diameters.

WORKED EXAMPLE

Figure 18.4 shows how the forces on two charged objects can be investigated. The two conducting spheres are identical, each having a diameter of 1.2 cm. They are charged by connecting to the same very high voltage supply. Use the data in Figure 18.4 to find the charges on the two spheres.

$(\varepsilon_0 = 8.85 \times 10^{-12} \, \text{C}^2 \text{N}^{-1} \text{m}^{-2})$

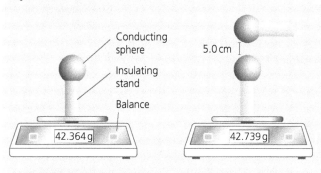

▲ Figure 18.4 Investigation of the forces between two charged conductors

Answer

distance between centres of the spheres = 5.0 +
$$(\tfrac{1}{2} \times 2 \times 1.2) \, \text{cm} = 6.2 \times 10^{-2} \, \text{m}$$

difference in reading on the balance = 42.739 − 42.364
$$= 0.375 \, \text{g}$$

force between the spheres = $(0.375 \times 10^{-3}) \times 9.81$
$$= 3.68 \times 10^{-3} \, \text{N}$$

Using:

$$F = \frac{Q_1 Q_2}{4\pi\varepsilon_0 r^2}$$

$$3.68 \times 10^{-3} = \frac{Q^2}{4\pi \times (8.85 \times 10^{-12}) \times (6.2 \times 10^{-2})^2}$$

$$Q^2 = (3.68 \times 10^{-3}) \times 4\pi \times (8.85 \times 10^{-12}) \times (6.2 \times 10^{-2})^2$$
$$= 1.57 \times 10^{-15} \, \text{C}^2$$

$$Q = 4.0 \times 10^{-8} \, \text{C}$$

NOW TEST YOURSELF   TESTED ☐

5   A helium nucleus consists of two protons and two neutrons and has a diameter of approximately 10^{-15} m. The charge on a proton is 1.6×10^{-19} C. Estimate the force between the two protons.
6   Two electrons are separated by a distance of 2.0×10^{-8} m. Calculate the force between them.
    (charge on an electron = 1.6×10^{-19} C)

Electric field strength of a point charge

We have already defined electric field strength as the force per unit charge on a stationary positive point charge placed at a point (p. 148). This means that the electric field strength around a point charge in free space is given by the equation:

$$E = \frac{Q}{4\pi\varepsilon_0 r^2}$$

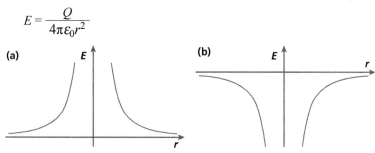

▲ Figure 18.5 The electric field near (a) a positive point charge and (b) a negative point charge

WORKED EXAMPLE

Calculate the electric field strength at a distance of 10 nm from a proton.

Answer

$$E = \frac{Q}{4\pi\varepsilon_0 r^2} = \frac{1.6 \times 10^{-19}}{4\pi \times (8.85 \times 10^{-12}) \times (10 \times 10^{-9})^2} = 1.4 \times 10^7 \, \text{N} \text{C}^{-1}$$

7   Calculate the electric field strength at the surface of the dome of a Van de Graaff generator of radius 30 cm that carries a charge of 800 nC.

Electric potential

Electric potential at a point

REVISED ☐

We saw in the work on gravity how gravitational potential at a point is defined as the work done in bringing unit mass from infinity to that point (p. 124). When considering the electric field, the rules are similar:

»   Choose a point that is defined as the zero of electric potential – infinity.
»   The electric potential at a point is then defined as the work done per unit positive charge in bringing a small test charge from infinity to that point.

Equation for the potential (*V*) near a point charge:

$$V = \frac{Q}{4\pi\varepsilon_0 r}$$

Note the similarity with gravitational potential. However, there are two types of electric charge: positive and negative, and consequently two types of field: repulsive and attractive.

Although this equation refers to a point charge, it is, as with the gravitational example, a good approximation provided that:

»   the charge is considered to be at the centre of the charged object and the distance is measured from this point
»   the point considered is at a distance greater than the radius of the charged object

It follows from this that the electric potential energy (E_p) of a charge Q_2 in an electric field is given by the equation:

$$E_p = \frac{Q_1 Q_2}{4\pi\varepsilon_0 r}$$

»   Consider the potential energy of a positive test charge brought up to a positively charged object and when it is brought up to a negatively charged object.
»   The test charge has zero potential energy at infinity.
»   When the **positive** test charge is brought up to the **positively** charged object, it **gains** potential energy (Figure 18.6a).
»   When the **positive** test charge is brought up to the **negatively** charged object, it **loses** energy, so it has a negative potential energy (Figure 18.6b).

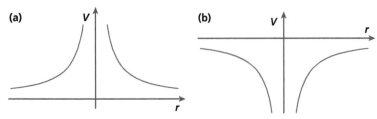

▲ Figure 18.6 (a) When a **positive** charge is brought towards a **positively** charged object it **gains** potential energy; a 'potential hill' is formed. (b) When a **positive** charge moves towards a **negatively** charged object, it **loses** potential energy; a 'potential well' is formed

Figure 18.7 should help you to understand why infinity is a good choice of position as the zero of potential energy.

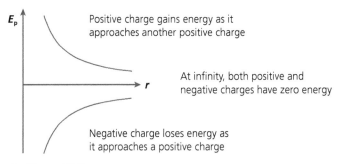

Positive charge gains energy as it approaches another positive charge

At infinity, both positive and negative charges have zero energy

Negative charge loses energy as it approaches a positive charge

▲ Figure 18.7

WORKED EXAMPLE

A proton travels directly towards the nucleus of an atom of silver at a speed of $5.00 \times 10^6\,\text{m s}^{-1}$. Calculate:

a the initial kinetic energy of the proton

b its closest approach to the silver nucleus

You may consider both the proton and the silver nucleus to be point charges.

(charge on the proton = +e; mass of a proton = $1.66 \times 10^{-27}\,\text{kg}$; charge on the silver nucleus = $+47e$, $e = 1.60 \times 10^{-19}\,\text{C}$, permittivity of free space = $8.85 \times 10^{-12}\,\text{F m}^{-1}$)

Answer

a $E_k = \tfrac{1}{2}mv^2 = 0.5 \times (1.66 \times 10^{-27}) \times (5.00 \times 10^6)^2$
$= 2.08 \times 10^{-14}\,\text{J}$

b Assume that all the potential energy of the proton is converted into electric potential energy as it approaches the silver nucleus.

$$E_p = \frac{Q_1 Q_2}{4\pi\varepsilon_0 r} = \frac{(1.6 \times 10^{-19}) \times 47 \times (1.6 \times 10^{-19})}{4\pi \times (8.85 \times 10^{-12}) \times r}$$

$$= \frac{1.08 \times 10^{-26}}{r}$$

Equating this with the initial kinetic energy gives:

$$r = \frac{1.08 \times 10^{-26}}{2.08 \times 10^{-14}} = 5.20 \times 10^{-13}\,\text{m}$$

> ## NOW TEST YOURSELF
> TESTED ☐
>
> 8 Calculate the potential at the surface of the dome in question 7.
> 9 A hydrogen atom consists of a proton and an electron. The energy needed to totally remove the electron from the proton (i.e. ionise the atom) is $2.2 \times 10^{-18}\,\text{J}$. Calculate the distance between the electron and the proton.

Relationship between electric field strength and potential

REVISED ☐

When studying the uniform field, we saw that the electric field strength between two plates can be calculated from either of the following two equations:

$$E = \frac{F}{q} \text{ or } E = \frac{V}{d}$$

Figure 18.8 shows that in a uniform field, the potential changes linearly with the distance d moved between the plates.

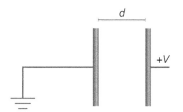
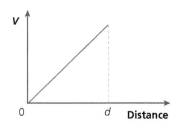

▲ Figure 18.8 The potential changes linearly in a uniform field

The second equation is only valid because the field between the plates is uniform and, hence, there is a linear change in potential from the earthed plate to the positive plate.

The electric field strength of a point charge gets weaker moving away from the charge. Consequently, the change in potential is not uniform (see Figure 18.9). Nevertheless, the change in potential with respect to distance is equal to the gradient of the graph:

$E = -$gradient of the V–r graph

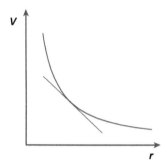

▲ Figure 18.9 The electric field strength at any point is equal to minus the gradient at that point, which is equal to $-(\Delta V/\Delta r)$

The relationship between the gradient of the potential curve $(\Delta V/\Delta r)$ and the electric field (E) is negative.

Electric potential energy

As in the case of gravity, two charges close to each other means that the system has electric potential energy. The formula for the electric potential energy of the system comes from the potential at a point multiplied by the charge on the second point charge:

$E_p = Qq/4\pi\varepsilon_0 r$

▶ REVISION ACTIVITIES

The similarity of the mathematics between gravitational fields and electric fields has already been referred to. Go back through both types of fields and note the similarities and differences to give you an in-depth understanding of the two situations. Take particular note of the differences in the mathematics of repulsion and attraction.

Must learn equations:

$F = qE$

$E = \dfrac{\Delta V}{\Delta d}$

$F = \dfrac{Q_1 Q_2}{4\pi\varepsilon_0 r^2}$

$E = \dfrac{Q}{4\pi\varepsilon_0 r^2}$

$V = \dfrac{Q}{4\pi\varepsilon_0 r}$

$E_p = \dfrac{Qq}{4\pi\varepsilon_0 r}$

Check your answers at www.hoddereducation.com/cambridgeextras

> ## END OF CHAPTER CHECK

In this chapter, you have learnt to:

» understand that an electric field is an example of a field of force ☐

» define electric field strength as force per unit charge on a stationary positive charge ☐

» recall and use $F = qE$ for the force on a charge in an electric field ☐

» represent an electric field by means of field lines ☐

» recall and use $E = \Delta V/\Delta d$ to calculate the magnitude of the field strength of the uniform field between parallel charged plates ☐

» describe the effect of a uniform field on the motion of charged particles ☐

» understand that, for a point outside a spherical conductor, the charge on a sphere may be considered to be at the centre of the sphere ☐

» recall and use Coulomb's law $F = Q_1Q_2/4\pi\varepsilon_0 r^2$ for the force between two point charges in free space ☐

» recall and use $E = Q/4\pi\varepsilon_0 r^2$ for the electric field strength around a point charge in free space ☐

» define electric potential at a point as the work done per unit positive charge in bringing a small test charge from infinity to that point ☐

» recall and use that the electric field strength at a point is equal to the negative of the potential gradient at that point ☐

» use $V = Q/4\pi\varepsilon_0 r$ for the electric potential in the field due to a point charge ☐

» recognise that the potential near a positive charge is greater than zero and is positive ☐

» recognise that the potential near a negative charge is less than zero and is negative ☐

» understand that the concept of electric potential leads to the electric potential energy of two point charges and use $E_p = Qq/4\pi\varepsilon_0 r$ ☐

19 Capacitance

Capacitors and capacitance

» Capacitors are electronic devices that store charge by separating charge.
» They can be considered to be made up of a pair of parallel conducting plates separated by an insulating material.
» When connected to a battery, charge flows onto one plate and an equal charge flows off the other plate (Figure 19.1).

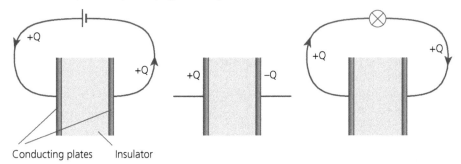

Conducting plates   Insulator

▲ **Figure 19.1 The flow of charge as a capacitor is charged and discharged**

» When the battery is removed, there is a net positive charge on one plate and a net negative charge on the other.
» When the capacitor is connected to a light bulb, the bulb lights up for a short time as the charge flows off the plates and round the circuit.

Uses of capacitors

Capacitors have various uses in circuits.

» In computers, they run the computer for a long enough time to save data if there is a power cut.
» They stop surges and sparking when high voltages are switched.
» They are used as a time delay. For example, in the timer on a burglar alarm system, which allows the operator time to leave the premises before the alarm switches on.

How much charge does a capacitor store?

This depends on the particular capacitor and the potential difference across it. The charge stored is proportional to the potential difference across the capacitor.

$$Q \propto V$$

This can be rewritten as an equality:

$$Q = CV$$

where C is a constant called the **capacitance** of the capacitor.

So:

$$C = \frac{Q}{V}$$

The unit of capacitance is the **farad (F)**.

> **KEY TERMS**
>
> The **capacitance** of a parallel plate capacitor is the charge stored on one plate per unit potential difference between the plates of the capacitor.
>
> 1 **farad (F)** is the capacitance of a capacitor that has a potential difference of 1 volt across the plates when there is a charge of 1 coulomb on the plates. (1 farad = 1 coulomb per volt)

A capacitance of 1 F is huge. In general, the capacitance of capacitors in electronic circuits is measured in microfarads (μF) or picofarads (pF):

$1\,\mu F = 10^{-6}\,F$

$1\,pF = 10^{-12}\,F$

WORKED EXAMPLE

Calculate the charge stored when a 2200 μF capacitor is connected across a 9 V battery.

Answer

$Q = CV = (2200 \times 10^{-6}) \times 9 = 0.020\,C\ (20\,mC)$

> ## NOW TEST YOURSELF
> TESTED ☐
>
> 1 A capacitor of capacitance 220 μF has a potential difference across its plates of 12 V. Calculate the charge stored on its plates.
> 2 A capacitor stores a charge of 200 nC when there is a potential difference of 6.0 V across its plates. Calculate the capacitance of the capacitor. Give your answer in pF.

Capacitance of an isolated object

REVISED ☐

It is not only capacitors that have capacitance – any isolated conducting object has capacitance. Consider an isolated conducting sphere of radius r carrying charge Q.

The potential at the surface of the sphere, V (see p. 152) is:

$$V = \frac{Q}{4\pi\varepsilon_0 r}$$

But

$$C = \frac{Q}{V}$$

So cancelling and rearranging gives:

$$C = 4\pi\varepsilon_0 r$$

WORKED EXAMPLE

Consider the Earth to be an isolated conducting sphere and estimate the capacitance of the Earth. (radius of the Earth = 6.4×10^6 m, $\varepsilon_0 = 8.85 \times 10^{-12}$)

Answer

$C = 4\pi\varepsilon_0 r = 4 \times \pi \times (8.85 \times 10^{-12}) \times (6.4 \times 10^6) = 7.1 \times 10^{-4}\,F \approx 700\,\mu F$

This is surprisingly small and it shows just how big the unit 1 farad is. Remember that this is the capacitance of an isolated object. Practical capacitors have their plates very close together to increase their capacitance.

> ## NOW TEST YOURSELF
> TESTED ☐
>
> 3 A football has a diameter of approximately 30 cm. Calculate its capacitance.

Capacitors in parallel

Consider the three capacitors in Figure 19.2.

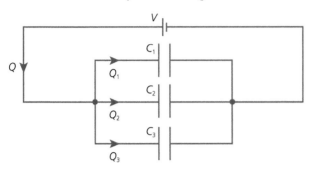

▲ Figure 19.2 Capacitors in parallel

From Kirchhoff's second law (p. 86), each capacitor has the same potential difference (V) across it.

From Kirchhoff's first law (p. 85), the total charge flowing from the cell is the sum of the charges on each of the capacitors:

$$Q = Q_1 + Q_2 + Q_3$$

From $Q = CV$:

$$C_{total}V = C_1V + C_2V + C_3V$$

The Vs can be cancelled, giving:

$$C_{total} = C_1 + C_2 + C_3$$

> **STUDY TIP**
> Compare this with the equation
> $R_{total} = R_1 + R_2 + R_3$
> which looks similar, but you must remember that whereas resistors in *series* add, capacitors in *parallel* add. Look at the capacitors in the parallel circuit. By adding a capacitor in parallel we are providing extra area for the charge to be stored on.

Capacitors in series

Consider three capacitors in series (Figure 19.3).

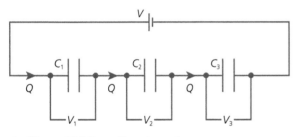

▲ Figure 19.3 Capacitors in series

From Kirchhoff's second law, the potential difference across the cell will equal the sum of the potential differences across the capacitors:

$$V = V_1 + V_2 + V_3$$

From $Q = CV$:

$$V = \frac{Q}{C}$$

where C is the capacitance of the circuit, so:

$$\frac{Q}{C_{total}} = \frac{Q_1}{C_1} + \frac{Q_2}{C_2} + \frac{Q_3}{C_3}$$

The same charge Q flows on to each capacitor. Therefore, the charges in the equation can be cancelled, so:

$$\frac{1}{C_{total}} = \frac{1}{C_1} + \frac{1}{C_2} + \frac{1}{C_3}$$

Check your answers at **www.hoddereducation.com/cambridgeextras**

This equation is of the same form as the formula for resistors in parallel. The combined resistance of several resistors in parallel is smaller than that of any of the individual resistors. Likewise, the combined capacitance of several capacitors in series is smaller than that of any of the individual capacitors.

WORKED EXAMPLE

A student has three capacitors of values 47 μF, 100 μF and 220 μF.

a Calculate the total capacitance when:
   i   all three capacitors are connected in parallel
   ii all three capacitors are connected in series
b How would the capacitors have to be connected to obtain a capacitance of 41 μF?

Answer

a i For capacitors in parallel:

   $$C_{total} = C_1 + C_2 + C_3 = 47 + 100 + 220 = 367\,\mu F$$

   ii For capacitors in series:

   $$\frac{1}{C_{total}} = \frac{1}{C_1} + \frac{1}{C_2} + \frac{1}{C_3} = \frac{1}{47} + \frac{1}{100} + \frac{1}{220} = 0.036\,\mu F^{-1}$$

   Therefore, $C = 28\,\mu F$

b The capacitance is less than 47 μF, the value of the smallest of the three capacitances.

Therefore, it is likely that the arrangement is of the form shown in Figure 19.4.

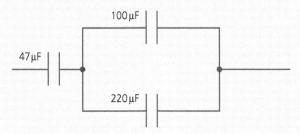

▲ **Figure 19.4**

Checking:

capacitance of the two capacitors in parallel =
$$100 + 220 = 320\,\mu F$$

These are in series with the 47 μF capacitor so:
$$\frac{1}{C_{total}} = \frac{1}{47} + \frac{1}{320} = 0.0244\,\mu F^{-1}$$

Therefore, $C = 41\,\mu F$

NOW TEST YOURSELF

TESTED ☐

4 Capacitors of 100 μF, 220 μF and 500 μF are connected in series. Calculate their combined capacitance.
5 A student has four 100 μF capacitors. How can he connect them in order to make a network with a capacitance of 133 μF?
6 Three capacitors have capacitance of 1500 μF, 2200 μF and 4700 μF. Calculate the total capacitance when they are connected:
   a in series
   b in parallel

Energy stored in a capacitor

Up to now we have described a capacitor as a charge store. It is more accurate to describe it as an energy store. The net charge on a capacitor is in fact zero: $+Q$ on one plate, $-Q$ on the other.

The energy stored in a capacitor is equal to the work done in charging the capacitor:

$W = \frac{1}{2}QV$

The half comes in because:

» when the first charge flows onto the capacitor plates there is no potential difference opposing the flow
» as more charge flows, the potential difference increases, so more work is done
» the average potential difference is equal to half the maximum potential difference

Using a potential–charge graph

>> Each strip in Figure 19.5 shows the work done on adding a small amount of charge to a capacitor.
>> It can be seen that the total work done in charging the capacitor, and hence the energy it stores, is equal to the area under the graph.
>> Compare this with the calculation of distance travelled by an object accelerating from rest on p.19.

It is sometimes useful to express the energy equation in terms of capacitance and potential difference. Substitute for Q using the basic capacitor equation, $Q = CV$:

$$W = \tfrac{1}{2}QV = \tfrac{1}{2}(CV)V$$

Therefore:

$$W = \tfrac{1}{2}CV^2$$

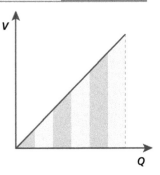

▲ Figure 19.5 Energy stored in a capacitor is equal to the area under the V–Q graph

WORKED EXAMPLE

1 Calculate the energy stored in a 470 µF capacitor when it is charged with a 12 V battery.

Answer

$W = \tfrac{1}{2}CV^2 = 0.5 \times (470 \times 10^{-6}) \times 12^2 = 3.4 \times 10^{-2}\,\text{J}$

2 A 50 µF capacitor is charged using a 12 V battery. It is then disconnected from the battery and connected across a second, uncharged 50 µF capacitor. Calculate:
   a the charge on the first capacitor before it is connected to the second capacitor
   b the energy stored on the first capacitor
   c the charge on each capacitor when they are connected together
   d the potential difference across the capacitors
   e the total energy stored when the two capacitors are connected together

Answer

a $Q = CV = 50 \times 12 = 600\,\mu\text{C} = 6.0 \times 10^{-4}\,\text{C}$
b $W = \tfrac{1}{2}QV = 0.5 \times (6 \times 10^{-4}) \times 12 = 3.6 \times 10^{-3}\,\text{J}$
c The charge will be shared equally between the two capacitors; therefore, each capacitor has a charge of $3.0 \times 10^{-4}\,\text{C}$.
d The charge on each capacitor is half the original value, so the p.d. will be one-half of the original = 6.0 V.
e Energy on one of the capacitors = $\tfrac{1}{2}QV = 0.5 \times (3.0 \times 10^{-4}) \times 6 = 9.0 \times 10^{-4}\,\text{J}$
   Therefore, the total energy = $2 \times 9.0 \times 10^{-4} = 1.8 \times 10^{-3}\,\text{J}$

► NOW TEST YOURSELF

TESTED

7 A 1500 µF capacitor has a potential difference of 9.0 V across its plates. Calculate:
   a the charge on the capacitor
   b the energy stored by the capacitor

Discharging a capacitor

A resistor is connected across a charged capacitor as in Figure 19.6.

» Switch S in the circuit in Figure 19.6 is closed.
» The potential difference across the capacitor drives a current through resistor R.
» The rate of discharge is not constant.
» As charge flows off the plates, the potential difference decreases.
» As a result, there is less potential difference driving the charge through the resistor, so the current is less.
» This type of decrease is common in nature and is referred to as an exponential decay. (See radioactive decay on pp. 198–200 and attenuation on pp. 205 and 209.)
» The graph of this decay is shown in Figure 19.7.

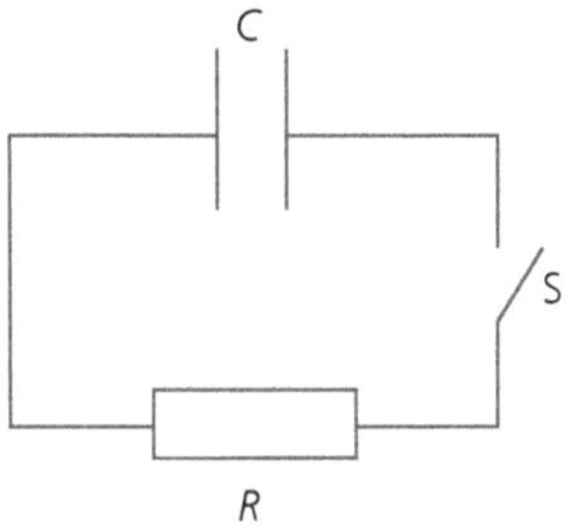

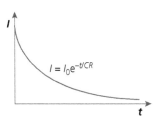

▲ Figure 19.6

STUDY TIP

One of the interesting features of the type of graph shown in Figure 19.7 is that the value reduces to the same fraction of the value in equal time intervals. For example, if the current halves in t seconds, then it will drop to half that value in a further t seconds, and halve again after another t seconds.

$$I = I_0 e^{-t/CR}$$

▲ Figure 19.7

The equation for this graph is:

$$I = I_0 \exp(-t/CR)$$

This is often written as:

$$I = I_0 e^{-t/CR}$$

where e is a naturally occurring number known as the exponential number (or sometimes Euler's number) = 2.71828..., C is the capacitance of the capacitor and R is the resistance of the resistor.

Three quantities with similar equations   REVISED ☐

» In the previous section, you saw that the discharge current of a capacitor was described by the equation:

$$I = I_0 e^{-t/CR}$$

» Now the current at any time $I = V/R$. R is a constant, so $I \propto V$ and therefore:

$$V = V_0 e^{-t/CR}$$

» The potential difference across the capacitor at any time $V = Q/C$. C is a constant, so $V \propto Q$ and thus:

$$Q = Q_0 e^{-t/CR}$$

Time constant of a discharge circuit   REVISED ☐

The initial current $I_0 = V_0/R$, where V_0 is the initial voltage across the capacitor.

But $V_0 = Q_0/C$. Therefore:

$$I_0 = Q_0/RC$$

» The larger R is, the smaller the initial current, and therefore the longer the capacitor takes to discharge.
» The larger C is, the more charge there is stored on the capacitor for a given p.d., thus the more charge needs to flow off the capacitor and the longer it takes to discharge.
» Consequently, RC is called the time constant (τ) of the circuit.
» Hence, $\tau = RC$.

WORKED EXAMPLE

A capacitor of capacitance $1000\,\mu F$ has a potential difference of $20\,V$ across it. It is connected across a resistor of resistance $400\,\Omega$. Calculate:

a   the initial charge on the capacitor

b   the time constant of the circuit

c   the time it takes for the potential difference across the capacitor to fall to $1.0\,V$

Answer

a $Q = CV = (1000 \times 10^{-6}) \times 20 = 2.0 \times 10^{-2}\,C = 0.020\,C$

b $\tau = RC = 400 \times (1000 \times 10^{-6}) = 4.0 \times 10^{-1}\,s = 0.40\,s$

c $V = V_0 \exp(-t/RC) \rightarrow 1.0 = 20 \exp(-t/0.40) \rightarrow$
$$\exp(-t/0.40) = 1/20 = 0.05$$

Taking natural logs of both sides of the equation:

$-t/0.40 = \ln(0.05) = -3.0$

Therefore:

$t = 3.0 \times 0.40 = 1.2\,s$

NOW TEST YOURSELF

TESTED ☐

8   Show that the unit of the time constant, RC, is seconds.

9   A capacitor is charged so that the potential difference across its plates is $20\,V$. The charge on the capacitor is $600\,mC$. The capacitor is discharged through a resistor of resistance $200\,\Omega$.

   a   Calculate the capacitance of the capacitor and the time it takes for half the charge to flow off it.

   b   State how much longer it would take for the charge stored to fall to one-quarter the original value.

REVISION ACTIVITIES

Look for appliances that contain capacitors and find out the capacitance of the capacitors in use.

'Must learn' equations:

$C = Q/V$

$C_{total} = C_1 + C_2 + C_3$ for capacitors in parallel

$\dfrac{1}{C_{total}} = \dfrac{1}{C_1} + \dfrac{1}{C_2} + \dfrac{1}{C_3}$ for capacitors in series

$W = \frac{1}{2}QV = \frac{1}{2}CV^2$

$I = I_0\exp(-t/RC),\ V = V_0\exp(-t/RC),\ Q = Q_0\exp(-t/RC)$

$\tau = RC$

END OF CHAPTER CHECK

In this chapter, you have learnt to:

» define capacitance as the charge stored per unit potential difference across a capacitor and around an isolated spherical conductor ☐

» recall and use the formula $C = Q/V$ ☐

» derive formulae for the combined capacitance of capacitors in parallel: $C_{total} = C_1 + C_2 + C_3$ ☐

» derive formulae for the combined capacitance of capacitors in series:
$1/C_{total} = 1/C_1 + 1/C_2 + 1/C_3$ ☐

» use the formulae for capacitors in parallel and in series ☐

» determine the electric potential energy stored in a capacitor from the area under the potential–charge graph ☐

» recall and use the formulae $W = \frac{1}{2}QV$ and $W = \frac{1}{2}CV^2$ ☐

» analyse graphs of the variation with time of potential difference, charge and current for a capacitor discharging through a resistor ☐

» use equations of the form $x = x_0 e^{-t/RC}$ where x could represent current, charge or potential difference for a capacitor discharging through a resistor ☐

» recall and use the formula $\tau = RC$ ☐

Check your answers at **www.hoddereducation.com/cambridgeextras**

Concepts of a magnetic field

» You should be familiar with magnetic fields from your earlier course.
» A magnetic field is, like gravitational fields and electric fields, a field of force.
» But there are significant differences between magnetic fields and electric and gravitational fields as well as similarities.

Magnetic effects

REVISED ☐

» Iron, cobalt and nickel and many of their alloys show **ferromagnetic** properties.
» Magnets are generally made of steel, an alloy of iron and carbon that retains magnetism much better than iron alone.
» The ends of a magnet are called poles.
» One end is called the north-seeking pole or N-pole and the other is called the south-seeking pole or S-pole.
» The laws of magnetism are that:
  » like poles repel
  » unlike poles attract

> **STUDY TIP**
>
> The term **ferromagnetic** compares the properties of cobalt and nickel with iron, the Latin name of which is ferrum.

Magnetic field shapes

REVISED ☐

» Like gravitational fields and electric fields, a magnetic field is a region in which a force is felt.
» The field shape can be shown using lines of magnetic force, which are known as **magnetic field lines**.
» In this case, when representing the field using lines of force, the lines of force represent the force on a single N-pole in the region around the field.
» Like electric and gravitational field diagrams, the stronger the field, the closer together the lines are drawn.

Figure 20.1 shows some magnetic fields.

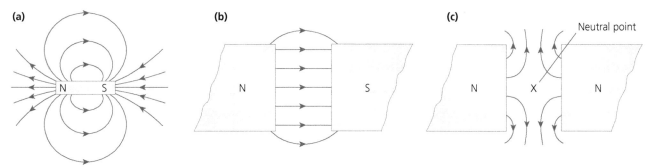

▲ Figure 20.1 (a) Magnetic field of a bar magnet, (b) magnetic field between opposite poles, (c) magnetic field between like poles; at the neutral point the two fields cancel out and there will be no force on a single north pole situated at this point

From Figure 20.1, you can see that the lines of magnetic field start at a north pole and finish at a south pole, and that they never cross or touch.

Magnetic fields due to currents

Magnetic field due to current in a straight conductor REVISED ☐

It is not only magnets that have an associated magnetic field – currents also do.

The magnetic field of a straight, current-carrying conductor can be investigated using the apparatus shown in Figure 20.2.

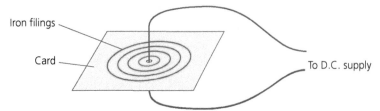

▲ **Figure 20.2 When iron filings are scattered near a current-carrying conductor, the iron filings line up, showing the existence of a magnetic field near the conductor**

» The field of a straight, current-carrying conductor is a set of concentric circles.
» The direction of the field depends on the current direction.

You can work out the direction by using the screw rule – imagine you are screwing a screw into the paper:

» The screwdriver turns clockwise, the same direction as the field lines for a current going into the plane of the paper.
» For a current coming out of the paper imagine you are loosening the screw. You must turn in an anticlockwise direction (Figure 20.3).

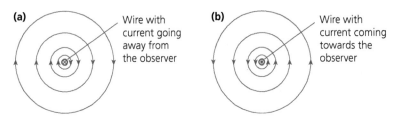

▲ **Figure 20.3**

The magnetic fields of two pairs of conductors with currents in the opposite and same directions are shown in Figure 20.4.

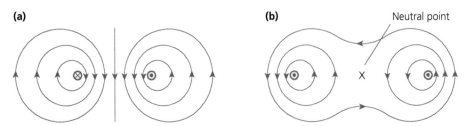

▲ **Figure 20.4 (a) The magnetic field of a pair of conductors with currents in opposite directions; note that this also shows the field of a narrow coil viewed at 90° to the axis of the coil. (b) The magnetic field of a pair of conductors with currents in the same direction; note the neutral point where the fields cancel each other out**

Check your answers at **www.hoddereducation.com/cambridgeextras**

A solenoid is a long coil. Its magnetic field is shown in Figure 20.5.

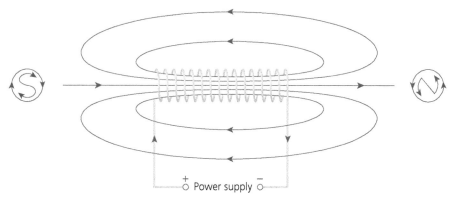

▲ Figure 20.5 The field around a current in a solenoid is similar to that of a bar magnet and is very strong

» The circles containing the N and S show a way of remembering which end of the solenoid acts as which pole.
» If you look at the right-hand end of the solenoid, the current direction is anticlockwise, the same as the arrows on the N.
» Look at the left-hand end. The current is in a clockwise direction, the same as the arrows on the S.

Solenoids are often used to make electromagnets, with the solenoid being wound round an iron or ferrous core. The presence of the ferrous core greatly increases the strength of the magnetic field inside (and near) the solenoid.

> **NOW TEST YOURSELF**   TESTED ☐

1 a Sketch the magnetic field between two S-poles.
   b State the difference between the magnetic field between two S-poles and the magnetic field between two N-poles.
2 a Sketch the magnetic field of a long, straight conductor carrying current into the plane of the page.
   b State how the field changes if the current is coming out of the plane of the paper.

Comparison of forces in gravitational, electric and magnetic fields

REVISED ☐

Table 20.1 shows the differences and similarities between the three types of force fields we have met.

▼ Table 20.1

	Gravitational	Electric	Magnetic
Stationary mass	Attractive force parallel to the field	None	None
Moving mass	Attractive force parallel to the field	None	None
Stationary charge	None	Attractive or repulsive force depending on type of charge, parallel to the field	None
Moving charge and electric current	None	Attractive or repulsive force depending on type of charge, parallel to the field	Force at right angles to both the field and the velocity of the charge/current. Force is a maximum when the velocity is at right angles to the field/current, and zero when the velocity/current is parallel to the field

Force on a current-carrying conductor

The motor effect: Fleming's left-hand rule

When a current-carrying conductor lies in a magnetic field, there is a force on the conductor. This is called the **motor effect** (Figure 20.6).

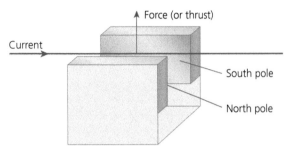

▲ **Figure 20.6 The force on a current-carrying conductor in a magnetic field**

» Observe that the current and the field are at right angles to each other and that the force is at right angles to both of these.
» This makes a set of three-dimensional axes.
» There will be no force on the conductor if it is parallel to the field.
» It requires, at least, a component of the current to be at right angles to it.

To help remember the specific directions of the different vectors we use **Fleming's left-hand rule** (Figure 20.7) in which:

» the **fi**rst finger represents the **fi**eld
» the **sec**ond finger represents the **c**urrent
» the **th**umb represents the **th**rust (or force)

▲ **Figure 20.7 Fleming's left-hand rule**

The magnitude of the force depends on:

» the magnitude of the current
» the length of conductor in the field
» the angle the conductor makes with the field (Figure 20.8)

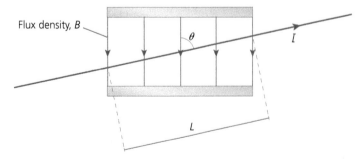

▲ **Figure 20.8 The force on a current-carrying conductor at an angle θ to a magnetic field**

A conductor of length L carrying a current I, lying at an angle θ to the field, will experience a force F vertically into the plane of the paper. Hence:

$$F \propto IL \sin\theta$$

Check your answers at **www.hoddereducation.com/cambridgeextras**

This can be written as:

$$F = BIL \sin \theta$$

where B is a constant, which can be considered as the **magnetic field strength**, although for historical reasons it is more usual to call it the **magnetic flux density**. Flux density is defined from the rearranged equation:

$$B = \frac{F}{IL \sin \theta}$$

The SI unit of magnetic flux density is the tesla (T), which is equivalent to $N\,A^{-1}\,m^{-1}$.

WORKED EXAMPLE

A copper power cable of diameter 2.5 cm carries a current of 200 A to a farm. There is a distance of 50 m between successive telegraph poles.

a Calculate the magnetic force on the section of the cable between two telegraph poles due to the Earth's magnetic field. (You may consider the wire to be at right angles to the Earth's magnetic field.)

b Compare this with the gravitational force on the cable.

(density of copper = 8900 kg m⁻³; flux density of the Earth's field = 30 μN A⁻¹ m⁻¹)

Answer

a $F = BIL \sin \theta = (30 \times 10^{-6}) \times 200 \times 50 \times \sin 90 = 0.30\,N$

b volume of the copper wire $= \pi r^2 L = \pi \times \left(\dfrac{2.5 \times 10^{-2}}{2}\right)^2 \times 50 = 0.025\,m^3$

mass = density × volume = 8900 × 0.025 = 220 kg

weight = mg = 220 × 9.81 ≈ 2200 N

This is almost four orders of magnitude (i.e. ×10 000) greater than the magnetic force on the cable.

NOW TEST YOURSELF

TESTED ☐

3 Figure 20.9 shows a conductor carrying a current perpendicular and into the plane of the page. There is a magnetic field going from left to right across the page. In which direction is the force on the conductor?

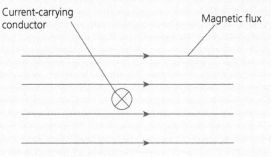

Current-carrying conductor

Magnetic flux

▲ **Figure 20.9**

4 A conductor carrying a current of 2.4 A is placed at right angles to a magnetic field of flux density 2.7 × 10⁻³ T. Calculate the force per unit length on the conductor.

5 A current-carrying conductor of length 25 cm at an angle of 60° to a magnetic field of flux density 4.8 × 10⁻³ T experiences a force of 3.3 × 10⁻³ N. Calculate the current in the conductor.

Forces on parallel, current-carrying conductors

You are already familiar with the fact that a current has an associated magnetic field (see pp. 164–165).

Figure 20.10 shows two parallel conductors with currents in opposite directions. It is a model in which the fields are considered separately. In practice, the two fields will combine.

>> Consider the effect of the field of the lower conductor on the upper conductor.
>> The magnetic field of the lower conductor is to the right of the page and the current in the upper conductor is vertically upwards, out of the plane of the paper.
>> If you apply Fleming's left-hand rule, you will see that there will be a force on the upper conductor away from the lower conductor.
>> If you consider the effect of the upper conductor on the lower conductor, you will see that the force is vertically down the page.
>> The two conductors repel.
>> A similar analysis shows that if the currents are both in the same direction, then the two conductors will attract.

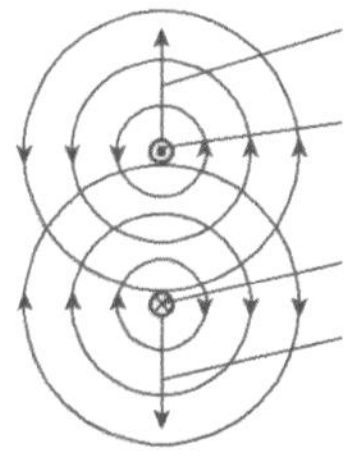

▲ **Figure 20.10 Fields of two current-carrying parallel conductors superimposed on each other.**

Force on a moving charge

Forces on charged particles in a magnetic field

>> Electric current is a flow of electric charge, so the magnetic force on a current is the sum of the forces on all the moving charge carriers that make up the current.
>> Alternatively, we could think of a beam of charged particles as a current.
>> Either way, we can deduce that the force on a charge q moving through a field of flux density B at speed v is given by the formula:

$F = Bqv \sin \theta$

where θ is the angle the velocity makes with the field.
>> The direction of the force on the particle can be found by using Fleming's left-hand rule. Remember that the second finger shows the direction of the conventional current.
>> Thus, the thumb shows the force direction on a positive charge.
>> A negative charge will experience a force in the opposite direction.
>> Study Figure 20.11. The force is at right angles to both the field and the velocity.
>> As the velocity changes, so does the direction of the force. Consequently, the particle travels in a circular path, with the magnetic force providing the centripetal force.

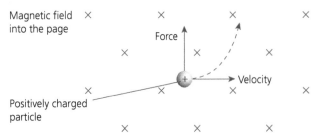

▲ **Figure 20.11 The path of a charged particle travelling with a velocity at right angles to the magnetic field**

You will study the tracks of charged particles in more detail in the section on charged particles (pp. 172–173).

Check your answers at **www.hoddereducation.com/cambridgeextras**

WORKED EXAMPLE

An electron travels with a constant velocity of $2.0 \times 10^5 \, \text{m s}^{-1}$. It enters a uniform magnetic field with flux density $25 \, \mu\text{N A}^{-1} \text{m}^{-1}$ at right angles to its direction of travel.

a Calculate the force on the electron.

b State the difference in the deflection if a proton, travelling at the same speed, enters the magnetic field at right angles.

Answer

a $F = Bqv \sin \theta$

$\theta = 90°$, therefore $\sin \theta = 1$

$F = Bqv = (25 \times 10^{-6}) \times (1.6 \times 10^{-19}) \times (2.0 \times 10^5) = 8.0 \times 10^{-19} \, \text{N}$

b The deflection would be in the opposite direction because the proton has positive charge, whereas the electron is negatively charged. The magnitude of the deflection would be much less because the proton has a much larger mass than the electron.

The Hall effect and Hall probe

REVISED ☐

Figure 20.12 shows a thin slice of a semiconductor.

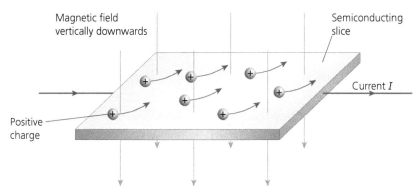

▲ **Figure 20.12**

» In this example, the current is carried by positive charge carriers.
» When a magnetic field is applied, the charge carriers are pushed to the rear side of the slice.
» This produces an e.m.f. between the front and rear edges of the slice, known as the Hall voltage (V_H).

Consider a slice of conductor of thickness t and width w carrying a current I at right angles to a magnetic field of flux density B (Figure 20.13).

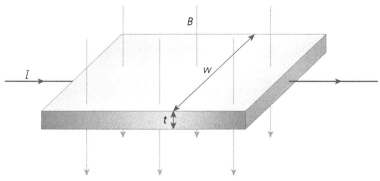

▲ **Figure 20.13**

» Each charge carrier has a charge q.
» The force on each charge carrier $F_{magnetic} = Bqv$, where v is the velocity of the charge carriers.
» This force tends to push the charge carriers across the slice, setting up a voltage V_H across the slice.
» The electric field produced by the migrating charge is:

$$E = \frac{V_H}{w}$$

» The electric force on each charge is:

$$F_{electric} = q\frac{V_H}{w}$$

» As charge builds up, it will limit further charge moving across the slice. Equilibrium will be reached when the magnetic force ($F_{magnetic}$) is equal to this electric force ($F_{electric}$):

So:

$$F_{electric} = q\frac{V_H}{w} = F_{magnetic} = Bqv$$

And:

$$\frac{V_H}{w} = Bv$$

» Now, $I = nAvq$ (see p. 77). Substituting:

$$v = \frac{I}{nAq}$$

in the previous equation gives:

$$\frac{V_H}{w} = \frac{BI}{nAq}$$

» A is the cross-sectional area $= wt$. Therefore:

$$\frac{V_H}{w} = \frac{BI}{nwtq}$$

So:

$$V_H = \frac{BI}{ntq}$$

» The variation in Hall voltage with flux density means the effect can be used to measure magnetic flux density. When used in this way it is known as a Hall probe.
» The Hall voltage produced in a metallic conductor is much less than in a semiconductor, so this device is always made from a semiconducting material.
» When using a Hall probe, care must be taken that the semiconductor slice is at right angles to the field being investigated.

> ### REVISION ACTIVITY
>
> Inspect the formula for the Hall voltage and explain why the Hall voltage for semiconductors (for the same current and flux density) is greater than for metals. (You should consider the numbers of charge carriers in both types of material and the effect on the drift velocity of those carriers to produce the same current.)

WORKED EXAMPLE

A strip of copper, 24 mm wide and 0.20 mm thick, is at right angles to a magnetic field of flux density 4.0×10^{-4} T. A current of 3.0 A passes through the copper strip. Calculate the Hall voltage produced across the width of the strip. (number of free electrons per unit volume $= 8.5 \times 10^{28}$ m^{-3})

Answer

$$V_H = \frac{BI}{ntq} = -\frac{(4.0 \times 10^{-4}) \times 3.0}{(8.5 \times 10^{28}) \times (2.0 \times 10^{-4}) \times (1.6 \times 10^{-19})} = 4.4 \times 10^{-10} \text{ V}$$

Check your answers at www.hoddereducation.com/cambridgeextras

> ### NOW TEST YOURSELF
> TESTED ☐
>
> 6 A Hall probe consisting of a semiconducting chip of thickness 0.12 mm and width 15 mm is used to measure the magnetic field between the jaws of a large horseshoe magnet. The current in the chip is 0.80 A and when the chip is at right angles to the magnetic field, the voltage across it is 50 mV. The number of charge carriers per unit volume (the number density) is $2.0 \times 10^{23}\,m^{-3}$. Calculate the flux density between the jaws of the magnet.

Deflection of charged particles in an electric field
REVISED ☐

Accelerating field

An electric field can be used to accelerate charged particles in the direction of the field. Early particle accelerators used electrostatic fields in this way. The kinetic energy given to the particle carrying charge q is:

$$E_k = Vq$$

Therefore:

$$\tfrac{1}{2}mv^2 = Vq$$

> ### WORKED EXAMPLE
>
> Calculate the voltage through which a proton of mass $1.66 \times 10^{-27}\,kg$, initially at rest, must be accelerated to reach a speed of $8 \times 10^6\,m\,s^{-1}$.
>
> **Answer**
>
> $$\tfrac{1}{2}mv^2 = Vq$$
>
> Rearrange the equation:
>
> $$V = \frac{\tfrac{1}{2}mv^2}{q} = -\frac{0.5 \times (1.66 \times 10^{-27}) \times (8.0 \times 10^6)^2}{1.6 \times 10^{-19}} = 3.3 \times 10^5\,V$$

Uniform electric field at right angles to the motion of the charged particle

Consider a positively charged particle moving at right angles to a uniform electric field (Figure 20.14).

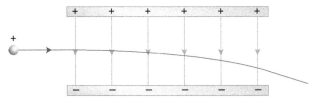

▲ Figure 20.14 A positively charged particle moving at right angles to a uniform electric field

➤➤ The force on the particle is vertically downwards.
➤➤ The path that the particle takes is similar to that of an object thrown in a uniform gravitational field (p. 25).
➤➤ There is no change to the horizontal component of the velocity but there is a constant acceleration vertically downwards. This produces the typical parabolic path.

WORKED EXAMPLE

Figure 20.15 shows the path of an electron travelling through a uniform electric field.

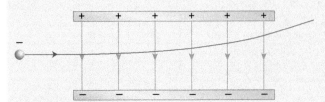

▲ **Figure 20.15**

On a copy of the diagram, sketch:

a the path that a positron moving at the same speed would take through the same field; label this A

b the path that a proton moving at the same speed would take through the same field; label this B

Answer

See Figure 20.16.

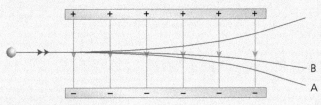

▲ **Figure 20.16**

> ### NOW TEST YOURSELF

TESTED ☐

7 An α-particle has a kinetic energy of 9.6×10^{-13} J. Calculate the potential difference it would need to be accelerated through to achieve this kinetic energy.

8 A proton has a speed of 3.8×10^{6} m s^{-1}. It enters a uniform magnetic field of flux density 2.8 mN A^{-1} m^{-1} at an angle of 40° to the field direction. Calculate the magnetic force on the proton.

STUDY TIP

The positron has the same mass as the electron and the same size charge, but the charge is positive. Consequently, it is deflected by the same amount as the electron but in the opposite direction. The proton has the same charge as the positron but is much more massive, so it will be deflected in the same direction as the positron but by much less. In practice, the deflection would be virtually undetectable.

Deflection of charged particles in a magnetic field

REVISED ☐

Measurement of *e/m*

» We saw on p. 168 that a particle of mass *m* carrying a charge *q* moving with a velocity *v* at right angles to a uniform magnetic field of flux density *B* experiences a force $F = Bqv$, and travels in a circular path.

» This force acts as the centripetal force, so:

$$Bqv = \frac{mv^2}{r}$$

Hence:

$$\frac{q}{m} = \frac{v}{Br}$$

» In Figure 20.17, the electron gun fires electrons vertically upwards.

» The field coils produce a uniform magnetic field into the plane of the paper, which causes the electrons to travel in a circular path.

» There is a trace of gas in the tube. When the electrons collide with the gas atoms, they cause ionisation (or excitation).

»» When the atoms drop back down to their ground state, they emit a pulse of light. In this way, the path of the electrons can be observed.
»» The energy of the electrons is calculated from the accelerating potential, and the diameter of their circular path is measured with a ruler.

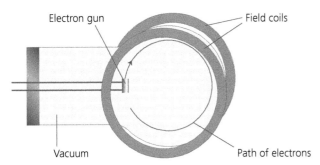

▲ Figure 20.17 Apparatus for measurement of e/m

WORKED EXAMPLE

Electrons travelling at a velocity of $4.0 \times 10^6 \,\mathrm{m\,s^{-1}}$ enter a uniform magnetic field of $0.60 \,\mathrm{mN\,A^{-1}\,m^{-1}}$ at right angles to the field. The electrons then travel in a circle of diameter 7.6 cm. Calculate the value of e/m for the electrons, and from this calculate the mass of an electron.

Answer

$$\text{radius of the circle} = \frac{7.6 \times 10^{-2}}{2} = 3.8 \times 10^{-2} \,\mathrm{m}$$

$$\frac{e}{m} = \frac{v}{Br} = \frac{4.0 \times 10^6}{(0.6 \times 10^{-3}) \times (3.8 \times 10^{-2})} = 1.75 \times 10^{11} \,\mathrm{C\,kg^{-1}}$$

The charge on the electron is 1.6×10^{-19} C and $e/m = 1.75 \times 10^{11} \,\mathrm{C\,kg^{-1}}$ hence:

$$m = \frac{e}{1.75 \times 10^{11}} = \frac{1.6 \times 10^{-19}}{1.75 \times 10^{11}} = 9.1 \times 10^{-31} \,\mathrm{kg}$$

> ## NOW TEST YOURSELF
> TESTED ☐
>
> 9 An α-particle has the same speed as the electron in the previous worked example and enters the same field as the electron. Calculate the diameter of the circle that the α-particle describes.

Velocity selectors

REVISED ☐

For more sophisticated measurement of q/m, the speed of the particles must be known much more precisely. One method of producing a beam of particles of all the same speed is to use crossed magnetic and electric fields (Figure 20.18).

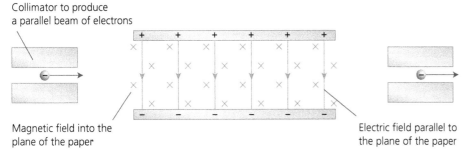

Magnetic field into the plane of the paper

Electric field parallel to the plane of the paper

▲ Figure 20.18 A velocity selector using crossed electric and magnetic fields

» The force on the electrons from the electric field is vertically upwards:

$F_E = eE$

» The force on the electrons from the magnetic field is vertically downwards:

$F_B = Bev$

» For electrons of a particular velocity, the two forces balance and cancel each other out. These electrons will be not be deflected and will pass straight through the apparatus and through the second collimator.
» Electrons of a slightly higher (or lower) speed will have a larger (or smaller) magnetic force on them and will be deflected and will not pass through the second collimator.
» For the selected velocity:

$F_E = F_B$

» Therefore:

$eE = Bev$

and thus:

$v = \dfrac{E}{B}$

» Although this equation has been worked out for electrons, neither the charge nor the mass of the particle feature in the final equation and hence it is valid for any charged particle.

WORKED EXAMPLE

In order to find the velocity of α-particles from a radioactive source, a narrow beam of the particles is incident on crossed electric and magnetic fields. The α-particles travel straight through the fields when the electric field strength is $8.6 \times 10^3\,V\,m^{-1}$ and the magnetic flux is $0.48\,mT$. Calculate the velocity of the α-particles.

Answer

$v = \dfrac{E}{B} = \dfrac{8.6 \times 10^3}{0.48 \times 10^{-3}} = 1.8 \times 10^7\,m\,s^{-1}$

REVISION ACTIVITY

Look up on the internet photographs of the tracks of fundamental particles produced in high-energy collisions. See if you can identify which particles have opposite charges, and those that have a very high mass.

NOW TEST YOURSELF

TESTED ☐

10 Protons of energy $1.2 \times 10^{-13}\,J$ enter crossed electric and magnetic fields. The field strength of the electric field is $12\,kV\,m^{-1}$. Calculate the flux density of the magnetic field if the protons are to move through the fields without deflection.

Electromagnetic induction

» Moving a wire perpendicularly to a magnetic field, as in Figure 20.19, induces an e.m.f. across the ends of the wire.

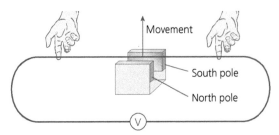

▲ Figure 20.19 Induction of an e.m.f.

» The magnitude of the e.m.f. is proportional to the magnetic field strength (the flux density), the length of wire in the field and the speed at which the wire is moved.

» When a second loop is made in the wire, the induced e.m.f. is doubled (Figure 20.20).

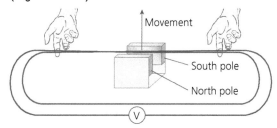

▲ Figure 20.20 Effect of two coils of wire on the induction of an e.m.f.

» Further loops show that the induced e.m.f. is proportional to the number of loops, N.

» This can be rewritten as:

$$E = -N\frac{\Delta\phi}{\Delta t}$$

where $\phi = BA$ and is called the magnetic flux, which is discussed in the next section. The minus sign shows that the induced e.m.f. is in such a direction to oppose the change that is causing it. See Lenz's law, p. 177.

This is often written:

$$E = -\frac{\Delta(N\phi)}{\Delta t}$$

Magnetic flux

REVISED ☐

A wire moving through a magnetic field sweeps out an area A, as shown in Figure 20.21.

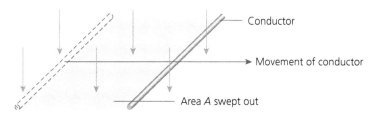

▲ Figure 20.21

The flux density multiplied by area (BA) is called the **magnetic flux** and has the symbol ϕ.

The unit of ϕ is $N\,m\,A^{-1}$, which is sometimes referred to as the weber (Wb).

Another useful term is $N\phi$, where N is the number of turns in a coil, called the **magnetic flux linkage.**

The induced e.m.f. is equal to the rate at which flux is swept out, often referred to as 'the rate of cutting flux'.

KEY TERMS

Magnetic flux is the product of magnetic flux density and the cross-sectional area perpendicular to the direction of the magnetic flux density.

Magnetic flux linkage is the product of the magnetic flux passing through a coil and the number of turns on the coil.

WORKED EXAMPLE

A small coil of cross-sectional area 2.4 cm² has 50 turns. It is placed in a magnetic field of flux density 4.0 mT, so that the flux is perpendicular to the plane of the coil. The coil is pulled out of the field in a time of 0.25 s. Calculate the average e.m.f. that is induced in the coil.

Answer

initial flux linkage $N\phi = NBA = 50 \times (4 \times 10^{-3}) \times$
$\qquad\qquad (2.4 \times 10^{-4}) = 4.8 \times 10^{-5}\,\text{Wb}$

$$E = -\frac{\Delta N\phi}{\Delta t} = -\frac{4.8 \times 10^{-5}}{0.25} = 1.9 \times 10^{-4}\,\text{V}$$

Laws relating to induced e.m.f.   REVISED ☐

Faraday's law of electromagnetic induction

The induced electromotive force (e.m.f.) is proportional to the rate of change of
magnetic flux linkage.

»» Notice that it is the *rate of change* of flux linkage, not just cutting through flux
   linkage, that induces an e.m.f.
»» A conductor in a magnetic field that changes has an e.m.f. induced across it, just
   as though the wire had been moved.
»» The conductor 'sees' a disappearing magnetic flux when the magnetic field is
   reduced and 'sees' a flux approaching when the field increases.

Experimental demonstration of electromagnetic induction

Set up the circuit as shown in Figure 20.22.

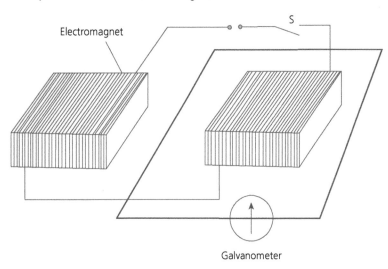

▲ **Figure 20.22**

Table 20.2 describes what happens when switch S is closed and opened, and what
happens when the conducting wire is moved within the jaws of the electromagnet.

▼ **Table 20.2**

Action	Observation
Close switch S	The needle on the galvanometer flicks to the right and goes back to the centre
Move the conducting wire vertically upwards between the jaws of the electromagnet	The needle on the galvanometer flicks to the left and goes back to the centre
Move the conducting wire vertically downwards between the jaws of the electromagnet	The needle on the galvanometer flicks to the right and goes back to the centre
Open switch S	The needle on the galvanometer flicks to the left and goes back to the centre

Check your answers at **www.hoddereducation.com/cambridgeextras**

You can see that switching the field on/off has exactly the same effect as putting the conductor into the field/taking it out of the field.

Lenz's law

» This law is really a statement of the conservation of energy.
» When a current is induced in a conductor, that current is in a magnetic field. Therefore, there is a force on it due to the motor effect.
» Work must be done against this force in order to drive the current through the circuit.

A formal statement of Lenz's law is as follows:

> The induced e.m.f. is always in such a direction so as to produce effects to oppose the change that is causing it.

Experimental demonstration of Lenz's law

» Figure 20.22 could form the basis of an experiment to demonstrate Lenz's law.
» A conducting disc is spun around so that it rotates freely. It will slow down gradually due to the frictional forces at the bearings.
» If the experiment is repeated so that the disc is between the jaws of a strong magnet, the disc will slow down much more quickly. The extra decelerating forces are produced by the currents induced in the disc.

▶ NOW TEST YOURSELF   TESTED ☐

12 A small coil of wire has 2000 turns and a cross-sectional area of $2.0\,\text{cm}^2$. It is placed perpendicular to a magnetic field of flux density $4.8 \times 10^{-3}\,\text{T}$.
   a Calculate:
      i   the flux through the coil
      ii the flux linkage in the coil
   b The coil is rotated so that its axis is at 45° to the field. Calculate the flux linkage in this position.
13 The coil of an electromagnet has 2000 turns and cross-sectional area $0.05\,\text{m}^2$. The coil carries a current that produces a magnetic field of flux density $5.6\,\text{mT}$. A switch is opened and the current and consequent magnetic field falls to zero in $120\,\mu\text{s}$. Calculate the e.m.f. induced across the terminals of the switch.

▶ REVISION ACTIVITY

'Must learn' equations:

$E = -\dfrac{\Delta(N\phi)}{\Delta t}$

$F = BIL\sin\theta$

$F = BQv\sin\theta$

$V_{\text{H}} = \dfrac{BI}{(ntq)}$

$\phi = BA$

END OF CHAPTER CHECK

In this chapter, you have learnt:

» that a magnetic field is an example of a field of force produced either by moving charges or by permanent magnets ☐

» to represent magnetic fields by field lines ☐

» to understand that a force might act on a current-carrying conductor placed in a magnetic field ☐

» to sketch magnetic field patterns due to the currents in a long, straight wire, a flat circular coil and a long solenoid ☐

» to understand that the magnetic field due to a current in a solenoid is increased by a ferrous core ☐

» to explain the origin of forces between current-carrying conductors and determine the direction of the forces ☐

» to recall and use the equation $F = BIL \sin\theta$, with directions as interpreted by Fleming's left-hand rule ☐

» to define magnetic flux density as the force acting per unit current per unit length on a conductor placed at right angles to the field ☐

» to determine the direction of the force on a charge moving in a magnetic field ☐

» to recall and use the equation $F = Bqv \sin\theta$ ☐

» to understand the origin of the Hall voltage and derive and use the expression $V_H = BI/(ntq)$, where t is the thickness ☐

» to understand the use of the Hall probe to measure magnetic flux density ☐

» to describe the motion of a charged particle in a uniform magnetic field perpendicular to the direction of movement of the particle ☐

» to explain how electric and magnetic fields can be used in velocity selection ☐

» to define magnetic flux as the product of the magnetic flux density and the cross-sectional area perpendicular to the direction of the magnetic flux density ☐

» to recall and use the equation $\Phi = BA$ ☐

» to understand and use the concept of magnetic flux linkage ☐

» to understand and explain experiments that demonstrate: ☐
  » that a changing magnetic flux can induce an e.m.f. in a circuit
  » the direction of the induced e.m.f. in a circuit
  » the factors effecting the magnitude of the induced e.m.f.

» to recall and use Faraday's and Lenz's laws of electromagnetic induction ☐

Check your answers at **www.hoddereducation.com/cambridgeextras**

21 Alternating currents

Characteristics of alternating currents

Terminology

Up to this point we have only looked at direct currents, which have been considered as steady, unchanging currents. An alternating current changes direction continuously – the charge carriers vibrate backwards and forwards in the circuit (Figure 21.1).

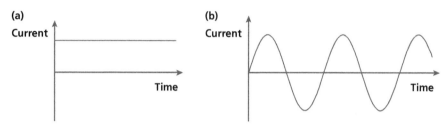

▲ Figure 21.1 (a) A direct current from a battery, (b) an alternating current from a simple generator

The terminology used in this section is similar to that used in the section on oscillations (see pp. 140–147).

» The **frequency** (f) of the alternating current is the number of complete cycles of the alternating current per unit time.
» The **period** (T) is the time taken for one complete cycle of the alternating current.
» The **peak current** (I_0) is the amplitude of the current during the cycle.

Although any current or potential difference which continually changes direction is an alternating current or potential difference, we shall concentrate on those which are described by an equation of the form $x = x_0 \sin \omega t$. This type of wave is known as sinusoidal.

You should recognise the shape of the curve from the work you have done on oscillations (see p. 141).

The equations for the current and the potential difference (voltage) are:

» for the current:

$I = I_0 \sin \omega t$

» for the voltage:

$V = V_0 \sin \omega t$

where I_0 is the peak current, V_0 is the peak voltage and ω is the angular frequency of the signal ($= 2\pi f$).

Power dissipated by an alternating current

REVISED ☐

Just as the current is changing continuously, the power is changing continuously.

A current I, which is equal to $I_0 \sin \omega t$, in a resistor of resistance R generates power just as a direct current does.

The power P generated by a current in the resistor is:

$P = I^2 R$

The power generated by an alternating current is:

$$P = (I_0 \sin \omega t)^2 R$$

$$P = I_0^2 R \sin^2 \omega t$$

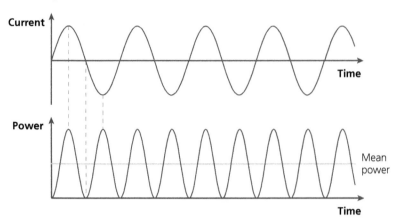

▲ Figure 21.2 Graphs of (a) the alternating current through a resistor and (b) the power dissipated in the resistor

» Study the power curve in Figure 21.2 – it is always positive.
» Even though the current goes negative, power is equal to current squared, and the square of a negative number is positive.
» The average (mean) power generated over a complete cycle is equal to half the peak power during that cycle (the horizontal line on the graph).

$$P_{\text{average}} = \tfrac{1}{2} P_0 = \tfrac{1}{2} I_0^2 R$$

The direct current that would dissipate this power $= \sqrt{\dfrac{1}{2} I_0^2}$ so:

$$I_{\text{d.c.}} = \frac{I_0}{\sqrt{2}}$$

This current is known as the **root-mean-square current** (see p. 136 for the r.m.s. velocities of particles in a gas).

$$I_{\text{r.m.s.}} = \frac{I_0}{\sqrt{2}}$$

Similarly, the r.m.s. voltage is given by:

$$V_{\text{r.m.s.}} = \frac{V_0}{\sqrt{2}}$$

(see p. 136 for the r.m.s.)

> **KEY TERMS**
>
> The **r.m.s. (root-mean-square) value** of the current (or voltage) is the value of direct current (or voltage) that would produce thermal energy at the same rate in a resistor.

WORKED EXAMPLE

a i Explain what is meant by the statement that a mains voltage is rated at 230 V, 50 Hz.
  ii Calculate the peak voltage.
b Calculate the energy dissipated when an electric fire of resistance 25 Ω is run from the supply for 1 hour.

Answer

a i 230 V tells us that the r.m.s. voltage is 230 volts. The frequency of the mains supply is 50 Hz.
  ii $V_0 = V_{\text{r.m.s.}} \times \sqrt{2} = 230 \times \sqrt{2} = 325\,\text{V}$
b energy $= \dfrac{V_{\text{r.m.s.}}^2 t}{R} = \dfrac{230^2 \times 60 \times 60}{25} = 7.6 \times 10^6\,\text{J}$

> ▶ **REVISION ACTIVITY**
>
> Ensure that you are fully aware that in this type of problem it is the r.m.s voltage that is used, not the peak voltage.

1   Calculate the r.m.s. current when the peak current is 2.4 A.
2   Calculate the peak voltage when the r.m.s. voltage is 48 V.
3   In an a.c. circuit, the peak potential difference across a resistor is 20 V and the peak current is 3 A. Calculate the mean power dissipated in the resistor.

Rectification and smoothing

Half-wave rectification   REVISED ☐

Although it is advantageous to transmit power using alternating currents, many electrical appliances require a direct current. The simplest way to convert an alternating supply to a direct supply is to use a single diode (Figure 21.3).

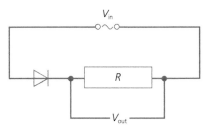

▲ Figure 21.3 Use of a diode as a rectifier

» The diode allows a current to pass only one way through it.
» The current through the resistor causes a potential drop across it when the diode conducts.
» When the input voltage is in the opposite direction, there is no current through the resistor, so the potential difference across it is zero.
» This is known as **half-wave rectification** (Figure 21.4).

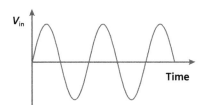

▲ Figure 21.4 Graph showing half-wave rectification

Full-wave rectification   REVISED ☐

With half-wave rectification, there is a current for only half a cycle. To achieve full-wave rectification, a diode bridge is used (Figure 21.5).

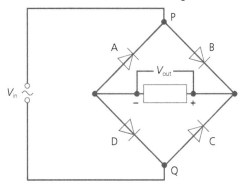

▲ Figure 21.5 Diode bridge for full-wave rectification

Study Figure 21.5.

When point P is positive with respect to Q:

» a current will pass from P through diode B
» through the resistor, then through diode D to point Q

When point P is negative with respect to Q:

» the current will pass from Q through diode C
» through the resistor, and through diode A to point P

In both cases, the current is in the same direction through the resistor, so the potential difference across it is always in the same direction. There is a full-wave rectified output voltage, as shown in Figure 21.6

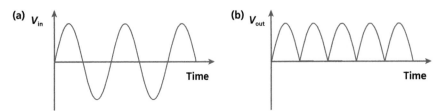

▲ Figure 21.6 (a) Input e.m.f. against time, (b) full-wave rectified output e.m.f.

> ## REVISION ACTIVITY

Copy out Figure 21.5 twice. On the first copy, consider point P to be positive and point Q to be negative. Draw arrows to show the currents in the diodes and through the resistor. On the second copy, repeat the exercise but this time consider point P to be negative and point Q to be positive. Draw arrows to show the currents in the diodes and through the resistor.

Note that the current (and hence the potential difference) across the resistor is in the same direction in both cases.

The output from a full-wave rectifier is still rough, rising from zero to a maximum and back to zero every half cycle of the original alternating input. Many devices, such as battery chargers, require a smoother direct current for effective operation. To achieve this, a capacitor is connected across the output resistor (Figure 21.7a).

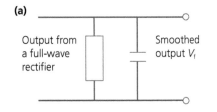 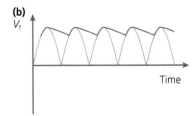

▲ Figure 21.7 Smoothing circuit and the smoothed output produced by using a single capacitor

» When the input voltage starts to fall from its peak, the capacitor starts to discharge.
» It takes some time to discharge, and will only partially discharge in the time it takes for the potential difference to rise once more.
» The value of the product RC (where C is the capacitance of the capacitor and R is the load resistance) is known as the time constant for the circuit (see p. 161).
» The value of RC *must* be much greater than the time period of the original alternating input.
» This means that the capacitor does not have sufficient time to discharge significantly.
» Figure 21.7(b) shows the smoothed output from a full-wave rectifier. The resultant voltage is not perfectly smooth – there is still a significant ripple.

Check your answers at **www.hoddereducation.com/cambridgeextras**

NOW TEST YOURSELF

TESTED ☐

4 The capacitance in a smoothing circuit is 2000 µF and the resistance is 150 Ω. Calculate the time constant of the circuit.

END OF CHAPTER CHECK

In this chapter, you learnt to:
» understand and use the terms period, frequency and peak value as applied to an alternating current or voltage ☐
» use equations of the form $x = x_0 \sin \omega t$ representing a sinusoidal alternating current or voltage ☐
» recall and use the fact that the mean power in a resistive load is half the maximum power for a sinusoidal alternating current ☐
» distinguish between the root-mean-square (r.m.s.) and peak values for a sinusoidal alternating current or voltage ☐
» recall and use $I_{r.m.s.} = I_0/\sqrt{2}$ and $V_{r.m.s.} = V_0/\sqrt{2}$ for a sinusoidal alternating current or voltage ☐
» distinguish graphically between half-wave and full-wave rectification ☐
» explain the use of a single diode for half-wave rectification ☐
» explain the use of four diodes for full-wave rectification ☐
» analyse the effect of a single capacitor in smoothing, including the effect of the value of capacitance and the load resistor ☐

REVISION ACTIVITY

'Must learn' equations:

$I = I_0 \sin \omega t$

$V = V_0 \sin \omega t$

$I_{r.m.s.} = \dfrac{I_0}{\sqrt{2}}$

$V_{r.m.s.} = \dfrac{I_0}{\sqrt{2}}$

22 Quantum physics

Energy of a photon

Waves or particles?

REVISED ☐

In the sections on waves and superposition, you learned that light shows properties of waves: polarisation (p. 63), diffraction (p. 71) and interference (p. 72). In this chapter, we investigate properties of light, which suggest that it also behaves like particles. You will also learn that electrons show wave properties.

It was Max Planck who first suggested that light might come in energy packets. Einstein linked this to his quantum theory. The resulting equation is known as the Einstein–Planck equation:

$$E = hf$$

where h is the Planck constant (6.63×10^{-34} J s).

These packets of energy are called photons. A photon is a quantum of electromagnetic energy.

> **WORKED EXAMPLE**
>
> An X-ray has a wavelength of 6.3×10^{-10} m. Calculate the energy of this X-ray photon.
>
> **Answer**
>
> Use the formula $c = f\lambda$ to find the frequency of the wave:
>
> $$f = \frac{c}{\lambda} = \frac{3.0 \times 10^8}{6.3 \times 10^{10}} = 4.8 \times 10^{17} \, \text{Hz}$$
>
> Use the formula $E = hf$ to find the energy of the photon:
>
> $$E = hf = (6.63 \times 10^{-34}) \times (4.8 \times 10^{17}) = 3.2 \times 10^{-16} \, \text{J}$$

It is often useful to give the Einstein–Planck equation in terms of wavelength rather than going through the whole process as in the previous worked example:

$$E = \frac{hc}{\lambda}$$

The electronvolt: a useful unit

REVISED ☐

When an electron is accelerated from rest through a voltage V, its kinetic energy E_k, measured in joules, is calculated by multiplying the charge on the electron by the accelerating voltage ($= eV$).

Rather than going through the process of using the formula and giving the energy in joules, we use the quantity eV itself as a unit of energy. We call this unit an **electronvolt (eV)**.

Thus, an electron accelerated through 1 V has an energy of 1 electronvolt. An electron accelerated through 50 000 V has an energy of 50 keV.

The charge on an electron $= 1.6 \times 10^{-19}$ C; consequently, a particle with an energy of 1 eV has an energy of 1.6×10^{-19} J.

A particle with energy 50 keV has an energy, in joules, of $50\,000 \times 1.6 \times 10^{-19} = 8.0 \times 10^{-15}$ J.

> **KEY TERMS**
>
> The **electronvolt (eV)** is a unit of energy equal to the energy gained by an electron when it is accelerated through a potential difference of 1 volt.

> **NOW TEST YOURSELF**　　　　　TESTED ☐

1 Calculate the energy of a photon of yellow light of frequency 5.1×10^{14} Hz. Give your answer in electronvolts and joules.
2 Calculate the energy of an X-ray photon of wavelength 4.0×10^{-10} m. Give your answer in electronvolts and joules.
3 An α-particle has an energy of 4.0 MeV.
   a Express its energy in joules.
   b What potential difference would it have to be accelerated through to reach this energy from rest?

> **STUDY TIP**
>
> The energy of any particle, including uncharged particles and photons, can be measured in electronvolts.
>
> But beware, an α-particle has a charge of magnitude $2e$, so when it is accelerated through 1 V, it has an energy of 2 eV.

Momentum of a photon

As well as a photon having a discrete energy and particle-type behaviour, it also has a momentum.

Einstein suggested that the momentum of a photon $p = E/c$, where E is the energy of the photon and c is the velocity of electromagnetic radiation in free space.

A useful way of using this formula is as $p = hf/c$, where h is the Planck constant and f is the frequency of the photon.

> **NOW TEST YOURSELF**　　　　　TESTED ☐

4 Calculate the momentum of an ultraviolet photon of wavelength 250 nm.

The photoelectric effect

The photoelectric effect can be demonstrated using a gold-leaf electroscope (Figure 22.1).

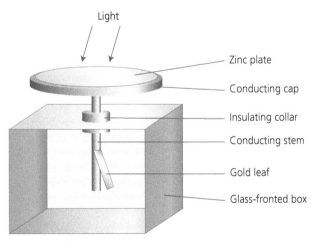

Light

— Zinc plate

— Conducting cap

— Insulating collar

— Conducting stem

— Gold leaf

— Glass-fronted box

▲ **Figure 22.1**

» The electroscope is charged negatively.
» When visible light is shone onto the zinc plate, the electroscope remains charged no matter how bright the light.
» When ultraviolet light is shone onto the plate, it steadily discharges; the brighter the light, the faster it discharges. Ultraviolet light has enough energy to lift electrons out of the plate and for them to leak away into the atmosphere; visible light does not have sufficient energy.

- ▶ This cannot be explained in terms of a wave model.
- ▶ If light is transferred by waves regardless of frequency, enough energy would arrive and electrons would escape from the metal surface.
- ▶ The packets of energy for visible light are too small to eject electrons from zinc.
- ▶ When higher frequency electromagnetic waves, such as ultraviolet light, fall on the zinc plate, electrons are emitted instantaneously.
- ▶ Electrons emitted in this way are often referred to as photoelectrons.
- ▶ Electromagnetic radiation arrives in packets of energy – the higher the frequency, the larger the packet. These packets of energy, or **quanta**, are called **photons**.
- ▶ The emission of photoelectrons occurs when a single photon interacts with an electron in the metal – hence, the instantaneous emission of the photoelectron.
- ▶ The ultraviolet packets are large enough to do this.
- ▶ More detailed experiments measuring the maximum energy of electrons emitted in the photoelectric effect (photoelectrons) give evidence for the relationship between photon energy and frequency.
- ▶ They also show that the intensity of the electromagnetic radiation does not affect the maximum kinetic energy with which the electrons are emitted, just the rate at which they are emitted (the photoelectric current).

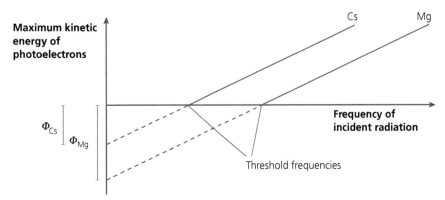

▲ Figure 22.2 Graph of maximum kinetic energy of photoelectrons against frequency of incident radiation

Figure 22.2 gives a great deal of information:

- ▶ The graphs are straight-line graphs with the same gradient, whatever metals are used.
- ▶ The **threshold frequency** and **threshold wavelength** for any metal is unique to that particular metal.
- ▶ The quantity Φ in the diagram is called the **work function energy**. This is dependent on the threshold frequency and is, therefore, different for each metal.
- ▶ The gradient is equal to the Planck constant.

KEY TERMS

The **threshold frequency** is the minimum-frequency radiation that is required to release electrons from the surface of a metal.

The **threshold wavelength** $= \dfrac{\text{speed of electromagnetic waves in free space}}{\text{threshold frequency}}$

The **work function energy** is the minimum energy, or minimum work required, to remove an electron from the surface of a metal.

The general equation for the graphs in Figure 22.2 is of the form:

$$y = mx + c$$

In this case:

$$E = hf - \Phi$$

where E is the maximum kinetic energy of a liberated electron.

Check your answers at **www.hoddereducation.com/cambridgeextras**

» When photoelectrons are ejected from the surface of a metal, the 'spare' energy of the photon (the energy that is not used in doing work to lift the electron out of the metal) is given to the electron as kinetic energy.

» When light is incident on the metal surface, electrons are emitted with a range of kinetic energies, depending on how 'close' they were to the surface when the photons were incident upon them.

» The maximum kinetic energy is when the minimum work is done in lifting the electron from the surface, consequently:

$hf = \Phi + \frac{1}{2}mv_{max}^2$

» It follows that $hf_0 = \Phi$, where f_0 is the threshold frequency of the radiation and the kinetic energy of the liberated electron is zero.

» This equation combines the wave nature (frequency) and the particle nature (energy of a photon) in the same equation.

» The equation also shows that the maximum energy of a photoelectron is independent of the intensity of the radiation and depends solely on the frequency of the radiation.

» Greater intensity means a greater rate of arrival of photons, but the energy of each photon is still the same. More electrons are liberated per unit time – but their maximum kinetic energy remains the same.

» The general equation for straight-line graphs is:

$y = mx + c$

» In this case:

$E = hf - \Phi$

where E is the maximum kinetic energy of a liberated electron.

» When photoelectrons are ejected from the surface of a metal, the 'spare' energy of the photon (the energy that is not used in doing work to lift the electron out of the metal) is given to the electron as kinetic energy.

» When light is incident on the metal surface, electrons are emitted with a range of kinetic energies, depending on how 'close' they were to the surface when the photons were incident upon them.

» The maximum kinetic energy is when the minimum work is done in lifting the electron from the surface, consequently:

$hf = \Phi + \frac{1}{2}mv_{max}^2$

» It follows that $hf_0 = \Phi$, where f_0 is the threshold frequency of the radiation and the kinetic energy of the liberated electron is zero.

> **STUDY TIP**
>
> The alternative way of expressing the energy of a photon, $E = \dfrac{hc}{\lambda}$, is derived from $E = hf$ and $f\lambda = c$. It is a useful expression to remember.

WORKED EXAMPLE

The work function energy of caesium is 2.1 eV.

a Calculate the threshold frequency for this metal.

b State in what range of the electromagnetic spectrum this radiation occurs.

c Radiation of frequency 9.0×10^{14} Hz falls on a caesium plate. Calculate the maximum speed at which a photoelectron can be emitted.
(mass of an electron = 9.1×10^{-31} kg; charge on an electron = 1.6×10^{-19} C, $h = 6.63 \times 10^{-34}$ J s)

Answer

a $\Phi = 2.1 \times (1.6 \times 10^{-19}) = 3.36 \times 10^{-19}$ J

$E = hf$

$f = \dfrac{E}{h} = -\dfrac{3.36 \times 10^{-19}}{6.63 \times 10^{-34}} = 5.1 \times 10^{14}$ Hz

b It is in the visible spectrum, in the yellow region.

c Energy of the photon:

$E = hf = (6.63 \times 10^{-34}) \times (9.0 \times 10^{14}) = 5.97 \times 10^{-19}\,\text{J}$

$hf = \Phi + E_k$

$E_k = hf - \Phi = (5.97 \times 10^{-19}) - (3.36 \times 10^{-19}) = 2.61 \times 10^{-19}\,\text{J}$

$E_k = \tfrac{1}{2}mv^2$

$v = \sqrt{\dfrac{2E_k}{m}} = \sqrt{\dfrac{2 \times (2.61 \times 10^{-19})}{9.1 \times 10^{-31}}} = 7.6 \times 10^5\,\text{m s}^{-1}$

NOW TEST YOURSELF

TESTED ▢

5 Calculate the minimum frequency of electromagnetic radiation that will liberate an electron from the surface of magnesium.
(work function energy of magnesium = 5.9×10^{-19} J)

6 Light of wavelength 4.5×10^{-7} m is incident on calcium. Calculate the maximum kinetic energy of an electron emitted from the surface.
(work function energy of calcium = 4.3×10^{-19} J)

Wave–particle duality

Electrons

REVISED ▢

'If light can behave like waves and like particles, can electrons behave like waves?'

That was the thought, in the 1920s, of postgraduate student Louis de Broglie, who first proposed the idea of 'matter waves'. If electrons do travel through space as waves, then they should show diffraction effects.

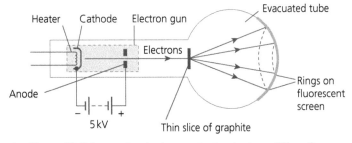

▲ Figure 22.3 Apparatus to demonstrate electron diffraction

» In the experiment in Figure 22.3, electrons are emitted from the hot cathode and accelerated towards the thin slice of graphite.
» The graphite causes diffraction and the maxima are seen as bright rings on the fluorescent screen.
» The diameter of the rings is a measure of the angle at which the maxima are formed.
» The diameters are dependent on the speed to which the electrons are accelerated.
» The greater the speed, the smaller the diameter, and hence the smaller the diffraction angle. From this information it can be concluded that:
  » electrons travel like waves
  » the wavelength of those waves is similar to the spacing of the atoms in graphite (otherwise diffraction would not be observed)
  » the wavelength of the waves decreases with increasing speed of the electrons

The de Broglie equation

It was proposed by de Broglie that the wavelength associated with electrons of mass m travelling at a velocity v could be determined from the formula:

$$\lambda = \frac{h}{mv}$$

You will recall that the quantity mv is the momentum p. The equation can be rewritten as:

$$\lambda = \frac{h}{p}$$

> **STUDY TIP**
>
> All matter has an associated wave function and it is the momentum, rather than the speed, that is the determining factor of the wavelength.

WORKED EXAMPLE

Electrons accelerated through a potential difference of 4.0 kV are incident on a thin slice of graphite that has planes of atoms 3.0×10^{-10} m apart. Show that the electrons would be suitable for investigating the structure of graphite.
(mass of an electron = 9.1×10^{-31} kg)

Answer

energy of the electrons = $4.0 \, \text{keV} = (4.0 \times 10^3) \times$
$$(1.6 \times 10^{-19}) = 6.4 \times 10^{-16} \, \text{J}$$

$E_k = \frac{1}{2}mv^2$

$$v = \sqrt{\frac{2E_k}{m}} = \sqrt{\frac{2 \times (6.4 \times 10^{-16})}{9.1 \times 10^{-31}}} = 3.8 \times 10^7 \, \text{m s}^{-1}$$

$$v = \sqrt{\frac{2E_k}{m}} = \sqrt{\frac{2 \times 6.4 \times 10^{-16}}{9.1 \times 10^{-31}}} = 3.8 \times 10^7 \, \text{m s}^{-1}$$

$$\lambda = \frac{h}{mv} = \frac{6.63 \times 10^{-34}}{(9.1 \times 10^{-31}) \times (3.8 \times 10^7)} = 1.9 \times 10^{-11} \, \text{m}$$

The waves are of a similar order of magnitude to the atomic layers of graphite and are therefore suitable for investigating the structure of graphite.

Other matter waves

» It is not just electrons that have an associated wave function; all matter does.
» Neutron diffraction is an important tool in the investigation of crystal structures because neutrons are uncharged.
» From the de Broglie equation, you can see that, for the same speed, neutrons will have a much shorter wavelength than electrons because their mass is much larger. Consequently, slow neutrons are used when investigating at the atomic level.

What about people-sized waves? Consider a golf ball of approximate mass 50 g being putted across a green at 3.0 m s^{-1}. Its wavelength can be calculated:

$$\lambda = \frac{h}{mv} = \frac{6.63 \times 10^{-34}}{0.05 \times 3.0} = 4.4 \times 10^{-33} \, \text{m}$$

This is not even 1 trillionth the diameter of an atomic nucleus. No wonder we do not observe the wave function associated with everyday-sized objects.

▶ NOW TEST YOURSELF

7 An electron is travelling at a speed of 3.2×10^6 m s^{-1}.
   a Calculate the wavelength of this electron.
   b State and explain how the wavelength of a neutron travelling at the same speed would differ from that of the electron.
   (mass of an electron = 9.1×10^{-31} kg; mass of a neutron = 1.67×10^{-27} kg)
8 A neutron is travelling at a speed of 3.0×10^6 m s^{-1}. Calculate the wavelength of the neutron.
   (mass of a neutron = 1.67×10^{-27} kg)

Energy levels in atoms and line spectra

Emission line spectra

» You learnt about the spectra of hot objects in the AS Level course (pp. 61–62), the visible spectrum being a continuous band of light with one colour merging gradually into the next.
» Such a spectrum is called a continuous spectrum (see Figure 22.4a).
» When a large potential difference is put across low-pressure gases, they emit a quite different spectrum, consisting of a series of bright lines on a dark background.
» This type of spectrum is called an **emission line spectrum** (see Figure 22.4b).

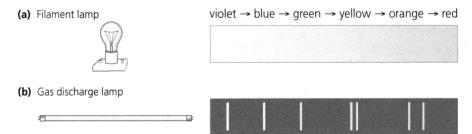

(a) Filament lamp

violet → blue → green → yellow → orange → red

(b) Gas discharge lamp

▲ Figure 22.4 (a) Continuous spectrum from a hot filament lamp, (b) emission line spectrum from a gas-discharge tube

» The precise lines visible depend on the gas in the discharge tube.
» Each gas has its own unique set of lines.
» This is useful for identifying the gases present, not only in Earth-based systems but in stars as well.
» Each line has a particular frequency. Hence, each photon from a particular line has the same energy.
» Atoms in gases are far apart from each other and have little influence on each other.
» The electrons in each atom can only exist in fixed energy states. Exciting the gas means that electrons are given energy to move from the lowest energy state (the ground state) to a higher energy state.
» They will remain in that state for a time before dropping back to a lower state. When they do so, they emit a photon.

Figure 22.5 shows the energy levels for a hydrogen atom and an electron falling from level $n = 3$ to $n = 1$.

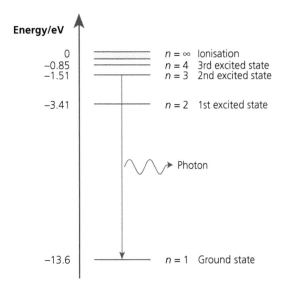

▲ Figure 22.5

STUDY TIP

When an electron is totally removed from the nucleus, the ionisation level is zero and the other energies are all negative. Compare this with the potential energies near a charged sphere.

Check your answers at **www.hoddereducation.com/cambridgeextras**

In general, the frequency of the emitted photon when an electron drops from a level E_1 to a level E_2 can be calculated from the equation:

$$hf = E_1 - E_2$$

WORKED EXAMPLE

Calculate the frequency of the photon emitted when the electron in the hydrogen atom in Figure 22.5 falls from the second excited state to the ground state.

Answer

$E = hf$

$E =$ the difference in the energy levels $= -1.51 - (-13.6) \approx 12.1\,\text{eV}$

$E = 12.1 \times (1.6 \times 10^{-19}) = 19.4 \times 10^{-19}\,\text{J}$

$f = \dfrac{E}{h} = -\dfrac{19.4 \times 10^{-19}}{6.63 \times 10^{-34}} = 2.9 \times 10^{15}\,\text{Hz}$

This is in the ultraviolet part of the spectrum.

» Emission line spectra give strong evidence for the existence of discrete energy levels in atoms.
» The photons emitted are of a definite set of frequencies and therefore of a definite set of energies.
» Each specific energy photon corresponds to the same fall in energy of an orbital electron as it drops from one discrete energy level to a second, lower, discrete energy level.

> ## NOW TEST YOURSELF
> TESTED ☐
>
> 9 a A hydrogen atom is in its first excited state and drops to the ground state. Using the information in Figure 22.5, calculate the wavelength of the photon that is emitted.
>   b In which part of the electromagnetic spectrum is the radiation from this transition?
>   c Suggest a transition that would produce a photon from the visible region of the spectrum.

Absorption line spectra

REVISED ☐

When white light from a continuous spectrum is shone through a gas or vapour, the spectrum observed is similar to a continuous spectrum, except that it is crossed by a series of dark lines (Figure 22.6).

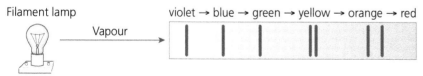

violet → blue → green → yellow → orange → red

Filament lamp

Vapour

▲ **Figure 22.6 An absorption line spectrum**

» This type of spectrum is called an **absorption line spectrum**.
» White light consists of all colours of the spectrum, which is a whole range of different frequencies and therefore photon energies.

» As the light goes through the gas/vapour, the photons of energy exactly equal to the differences between energy levels are absorbed, as shown for hydrogen in Figure 22.7.

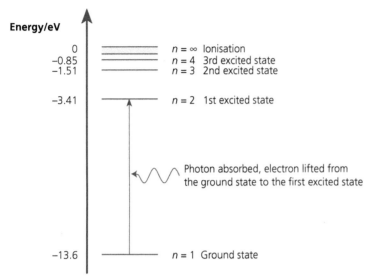

▲ Figure 22.7 Energy level diagram showing absorption of a photon, which excites an electron into a higher energy level

» The light is then re-emitted by the newly excited atom.
» However, this secondary photon can be emitted in any direction, so the energy of this frequency radiated towards the observer is very small.
» Consequently, dark lines are observed in the spectrum.

You may find that the absorption of energy from a photon of exactly the right frequency reminds you of resonance (p. 68).

WORKED EXAMPLE

The absorption line spectrum from a star is studied. A dark line is observed at a wavelength of 6.54×10^{-7} m. Calculate the difference in the two energy levels that produces this line.

Answer

$$E = hf = \frac{hc}{\lambda} = \frac{(6.63 \times 10^{-34}) \times (3 \times 10^{8})}{6.54 \times 10^{-7}} = 3.04 \times 10^{-19}\,\text{J}$$

STUDY TIP

If the answer to the worked example is converted to electronvolts, it becomes 1.90 eV. This is the difference between the first and second excited levels in the hydrogen atom. This reaction is quite likely – the outer atmosphere of the star, although cooler than the core, is still at a high temperature, so a lot of atoms will be in the first, and other, excited states. This provides evidence that the outer atmosphere of the star contains hydrogen, although more lines would need to be observed for this to be confirmed.

▶ REVISION ACTIVITY

'Must learn' equations:

$E = hf$

$p = \dfrac{E}{c} = \dfrac{hf}{c}$

$E = hf - \Phi$

$\lambda = \dfrac{h}{p}$

$hf = \Phi + \tfrac{1}{2}mv_{max}^2$

$hf = E_1 - E_2$

NOW TEST YOURSELF

TESTED ☐

10 Explain why the maximum kinetic energy of photoelectrons liberated from the surface of a metal is determined by the wavelength of the radiation incident on the metal.

END OF CHAPTER CHECK

In this chapter, you have learnt to:
» understand that electromagnetic radiation has a particulate nature ☐
» understand that a photon is a quantum of electromagnetic energy ☐
» recall and use $E = hf$ ☐
» use the electronvolt (eV) as a unit of energy ☐
» understand that a photon has momentum ☐
» recall that the momentum of a photon is given by $p = E/c$ ☐
» understand that photoelectrons may be emitted from a metal surface when it is illuminated by electromagnetic radiation ☐
» understand the use of the terms threshold frequency and threshold wavelength ☐
» explain photoelectric emission in terms of photon energy and work function energy ☐
» recall and use $hf = \Phi + \frac{1}{2}mv_{max}^2$ ☐
» explain why the maximum kinetic energy of photoelectrons is independent of intensity whereas the photoelectric current is proportional to intensity ☐
» understand that the photoelectric effect provides evidence for a particulate nature of electromagnetic radiation while phenomena such as interference and diffraction provide evidence for the wave nature ☐
» describe and interpret qualitatively the evidence provided by electron diffraction for the wave nature of particles ☐
» understand the de Broglie wavelength associated with a moving particle ☐
» recall and use $\lambda = h/p$ ☐
» understand that there are discrete electron energy levels in isolated atoms (e.g. atomic hydrogen) ☐
» understand the appearance and formation of emission and absorption line spectra ☐
» recall and use the relationship $hf = E_1 - E_2$ ☐

23 Nuclear physics

Mass defect and nuclear binding energy

Mass and energy

The mass of a proton = 1.673×10^{-27} kg, and the mass of a neutron = 1.675×10^{-27} kg. The mass of the carbon-12 nuclide, which is made up of six neutrons and six protons, is 19.921×10^{-27} kg.

Consider the mass of six free protons plus six free neutrons:

$$(6 \times 1.673 \times 10^{-27}\,\text{kg}) + (6 \times 1.675 \times 10^{-27}\,\text{kg}) = (10.035\,738 + 10.049\,574) \times 10^{-27}\,\text{kg}$$

$$= 20.088 \times 10^{-27}\,\text{kg}$$

This is 0.167×10^{-27} kg more than the mass of the carbon-12 nuclide.

Where has the mass gone?

» The answer lies in the fact that when the nucleons are bound within a nucleus, they are in a lower energy state than when they are free.
» This and other evidence led Einstein to recognise that energy and mass are interwoven.
» He developed the concept that energy and mass are not two separate entities but a single entity which we call **mass–energy** linked by the equation:

$E = mc^2$

where c is the speed of electromagnetic radiation in free space.

» Generally, we shall be concerned with changes in mass owing to changes in energy, when the equation becomes $\Delta E = \Delta mc^2$ where ΔE is the change in energy and Δm is the resulting change in mass.

> **STUDY TIP**
>
> The kilogram is very large compared with the masses of atoms and subatomic particles. A useful unit to use is the unified atomic mass unit (u).
>
> 1 u is equal to 1.660540×10^{-27} kg and is defined as being equal to one-twelfth of the mass of a carbon-12 atom.

> **STUDY TIP**
>
> The more energy a particle has, the more mass it has. It does not matter what type of energy it is, whether it is potential energy or kinetic energy.
>
> One of the major pieces of evidence for the increase in mass with an increase of energy is that the deflection of highly energetic charged particles (that is, charged particles travelling at very high speeds) was less than expected when in a magnetic field.
>
> Einstein suggested that one way of looking at the mass–energy concept is to consider that the energy itself has mass. In everyday experience, this extra mass is so small we do not notice it – it is only in extreme circumstances that it becomes significant.

> **WORKED EXAMPLE**
>
> A proton in a particle accelerator is accelerated through 4.5 GV. Calculate the increase in mass of the proton.
>
> **Answer**
>
> energy gained by the proton = 4.5 GeV
>
> Convert this into joules:
>
> $$E = (4.5 \times 10^9) \times (1.6 \times 10^{-19}) = 7.2 \times 10^{-10}\,\text{J}$$
>
> $E = mc^2$
>
> Therefore:
>
> $$m = \frac{E}{c^2} = \frac{7.2 \times 10^{-10}}{(3.0 \times 10^8)^2} = 8.0 \times 10^{-27}\,\text{kg}$$

This is an amazing result. The rest mass of a proton is 1.66×10^{-27} kg. The increase in mass is almost five times this, giving a total mass of almost six times the rest mass.

> **NOW TEST YOURSELF**   TESTED ☐
>
> 1 A β⁻-particle has an energy of 1.2 MeV. Calculate the extra mass the
>   particle gains and compare it with the rest mass of an electron.
>   (rest mass of an electron = 9.11×10^{-31} kg)

Mass defect and binding energy   REVISED ☐

We saw on p. 194, that the nucleus of a carbon-12 atom has a mass of
0.167×10^{-27} kg less than the total mass of six protons and six neutrons.
The difference in mass is known as the **mass defect**.

> **KEY TERMS**
>
> The **mass defect** of a nuclide is the difference between the mass of the nucleus of a
> nuclide and the total mass of the nucleons of that nuclide, when separated to infinity.

> **STUDY TIP**
>
> Be careful here – there
> is electrostatic repulsion
> between the protons
> inside the nucleus.
> The attractive forces that
> hold the nucleus together
> are *not* electrostatic in
> nature and they are very
> much larger.

Just as with the energy levels in the outer atom and with the electrical energy of a
negative particle near a positive charge, the field inside a nucleus is attractive.

The zero of energy, as in the gravitational and electrostatic field, is taken
as infinity and, therefore, the particles in the nucleus have negative energy.
To separate a nucleus into its constituent protons and neutrons, work must be done.
This work is called the **binding energy**.

» The binding energy is different for the nucleus of each different isotope.
» The binding energy per nucleon is a measure of its stability. The more energy
  needed to tear the nucleus apart, the less likely it is to be torn apart.
» A binding energy curve showing the general trend and specific important
  nuclides is shown in Figure 23.1

> **KEY TERMS**
>
> The **binding energy** of
> a nuclide is the work
> done or energy required
> to separate to infinity
> the nucleus into its
> constituent protons and
> neutrons.

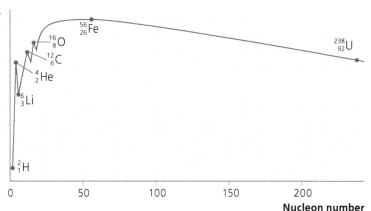

Binding energy per nucleon/ MeV (vertical axis)
Nuclides labelled: $^{56}_{26}$Fe, $^{16}_{8}$O, $^{12}_{6}$C, $^{4}_{2}$He, $^{6}_{3}$Li, $^{2}_{1}$H, $^{238}_{92}$U
Nucleon number (horizontal axis: 0, 50, 100, 150, 200)

▲ Figure 23.1

» In particular, note the high binding energies per nucleon for $^{4}_{2}$He, $^{12}_{6}$C and $^{16}_{8}$O.
» The highest binding energy per nucleon, and therefore the most stable nuclide,
  is $^{56}_{26}$Fe.
» Binding energies are very large and hence there is a measurable difference in the
  mass of a proton that is bound in a nucleus and that of a free proton at rest.
» The shape of the curve for the 'missing mass' per nucleon, known as the mass
  defect per nucleon, is the same as that for binding energy.

WORKED EXAMPLE

The helium nucleus $_2^4\mathrm{He}$ has a mass defect of 0.30629 u. Calculate:

a the mass defect in kilograms

b the binding energy

c the binding energy per nucleon
   (rest mass of a proton = 1.007276 u; rest mass of a neutron = 1.008665 u

Answer

a 1 u = 1.660540×10^{-27} kg
   Therefore:
   $0.30629\,\mathrm{u} = 0.30629 \times 1.660540 \times 10^{-27}$
   $= 5.09 \times 10^{-28}$ kg

b binding energy $E = mc^2 = 5.09 \times 10^{-28} \times (3.0 \times 10^8)^2$
   $\approx 4.58 \times 10^{-11}$ J

c binding energy per nucleon $= \dfrac{4.58 \times 10^{-11}}{4}$
   $= 1.15 \times 10^{-11}$ J

> ## NOW TEST YOURSELF
> TESTED ☐
>
> 2 The mass defect of a nuclide of the isotope $_6^{12}\mathrm{C}$ is 1.64×10^{-28} kg. Determine the binding energy per nucleon of this nuclide. Give your answer in MeV and in joules.

Nuclear reactions
REVISED ☐

You have already met radioactive decay earlier in the course (see pp. 95–96) and should be familiar with decay equations, such as that for americium decay when it emits an α-particle:

$$_{95}^{241}\mathrm{Am} \rightarrow {}_{93}^{237}\mathrm{Np} + {}_2^4\alpha + \text{energy}$$

Another common form of decay is β decay:

$$_{38}^{90}\mathrm{Sr} \rightarrow {}_{39}^{90}\mathrm{Y} + {}_{-1}^{0}\beta + \text{energy}$$

Although α, β and γ decay are the most common forms of decay, there are many other possibilities. An important example is the formation of carbon-14 in the atmosphere. A neutron is absorbed by a nitrogen nucleus, which then decays by emitting a proton:

$$_7^{14}\mathrm{N} + {}_0^1\mathrm{n} \rightarrow {}_6^{14}\mathrm{C} + {}_1^1\mathrm{p}$$

> **STUDY TIP**
>
> You should recognise that both the nucleon number and the proton number are conserved in every decay.

WORKED EXAMPLE

A $_8^{16}\mathrm{O}$ nucleus absorbs a neutron. The newly formed nucleus subsequently decays to form a $_9^{17}\mathrm{F}$ nucleus.

a Write an equation to show the change when the neutron is absorbed.

b Deduce what type of particle is emitted when the decay of the newly formed nucleus occurs.

Answer

a $_8^{16}\mathrm{O} + {}_0^1\mathrm{n} \rightarrow {}_8^{17}\mathrm{O}$

b The decay in the second part is described by an equation of the form:
$$_8^{17}\mathrm{O} \rightarrow {}_9^{17}\mathrm{F} + {}_{-1}^{0}Z$$
where Z is the unknown particle, which from its proton number and its nucleon number is a β-particle.

Note that throughout the reaction, the total nucleon number remains constant (at 17) and the proton number remains constant at 8.

Nuclear fission
REVISED ☐

» Fission is the splitting of a nucleus into two roughly equal-sized halves with the emission of spare neutrons.

» If you look at the binding energy curve in Figure 23.1, you will see that the nuclides with nucleon numbers between about 50 and 150 have significantly more binding energy per nucleon than the largest nuclides with nucleon numbers greater than 200.

Check your answers at **www.hoddereducation.com/cambridgeextras**

>> A few of these larger nuclides are liable to fission.
>> Fission happens in nature but it is fairly rare. However, physicists can induce fission by allowing large, more stable nuclides to capture a neutron to form an unstable nuclide.
>> For example, a uranium-235 nucleus, which is found in nature, can capture a slow-moving neutron to form a uranium-236 nucleus.

$$^{235}_{92}U + ^{1}_{0}n \rightarrow ^{236}_{92}U$$

>> This nucleus is unstable and will undergo fission (Figure 23.2):

$$^{236}_{92}U \rightarrow ^{146}_{57}La + ^{87}_{35}Br + 3^{1}_{0}n$$

In Figure 23.2:

>> Step 1 – a neutron moves slowly towards a U-235 nucleus.
>> Step 2 – the U-235 nucleus absorbs the neutron to form an unstable U-236 nucleus.
>> Step 3 – the U-236 nucleus splits into two roughly equal halves (the fission fragments), which fly apart; three neutrons are released, which also fly away at high speeds.
>> Most of the energy in fission is carried away by the fission fragments as kinetic energy, although some is carried away by the neutrons.
>> In addition, gamma rays are formed. Some are formed almost immediately and some are formed later as the nucleons in the fission fragments rearrange themselves into a lower, more stable energy state.
>> Fission is used in all working nuclear power stations.
>> If the fissionable nuclide being used is uranium, the neutrons released are slowed down, so that they cause new fissions to keep the process going.
>> For power generation, each fission must, on average, produce one new fission to keep the reaction going at a constant rate.
>> The earliest nuclear weapons used fission. The principle is the same, but in this case each fission must induce more than one new fission on average, so that the reaction rate rapidly gets faster and faster, causing an explosion.

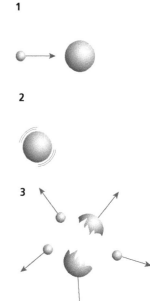

1

2

3

▲ Figure 23.2 A cartoon view of fission

> ## NOW TEST YOURSELF
> TESTED ☐
>
> 3 Copy and complete the equation, which shows a possible fission of plutonium.
>
> $$^{239}_{94}Pu + ^{1}_{0}n \rightarrow ^{...}_{...}Pu \rightarrow ^{150}_{...}Nd + ^{88}_{34}Se + ...^{1}_{0}n$$

Nuclear fusion
REVISED ☐

>> Nuclear fusion is, in many ways, the opposite of fission.
>> Two small nuclei move towards each other at great speeds, overcome the mutual electrostatic repulsion and merge to form a larger nucleus (Figure 23.3).
>> If you look at the binding energy curve in Figure 23.1 again, you will see that the binding energy per nucleon of deuterium ($^{2}_{1}H$) is much less than that of helium ($^{4}_{2}He$).
>> Two deuterium nuclei can fuse to form a helium nucleus with the release of energy. In practice, the fusion of tritium ($^{3}_{1}H$) and deuterium is more common:

$$^{2}_{1}H + ^{3}_{1}H \rightarrow ^{4}_{2}He + ^{1}_{0}n$$

>> Fusion releases much more energy per nucleon than fission.
>> The difficulty in using fusion commercially is that the pressure and temperature of the fusing mixture are extremely high – so high that all the electrons are stripped from the nuclei and the mixture becomes a sea of positive and negative particles called a plasma.

- » A physical container cannot be used to hold the plasma – it would immediately vaporise (not to mention cool the plasma).
- » Extremely strong magnetic fields are used to contain the plasma.
- » Even though scientists have not yet solved the problems of controlling nuclear fusion, we rely on it because it is the process that fuels the Sun (see p. 214).

In Figure 23.3:

- » Step 1 – a deuterium nucleus and a tritium nucleus move towards each other at high speed.
- » Step 2 – they collide and merge.
- » Step 3 – a helium nucleus and a neutron are formed and fly apart at high speed.

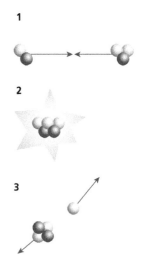

1

2

3

▲ Figure 23.3 A cartoon view of fusion

> ### NOW TEST YOURSELF
> TESTED ☐
>
> 4 The first stage of a fusion chain in the Sun is the fusion of two protons, resulting in the formation of deuterium ($^2_1$H) and the release of 3.27 MeV of energy.
> a Write an equation that describes this reaction.
> b Determine the mass defect per nucleon in the reaction.

Radioactive decay

- » Some nuclides are unstable and decay by emitting radiation; this is known as radioactive decay.
- » The rate of radioactive decay is not dependent on outside conditions (e.g. temperature, pressure).
- » In this sense, the decay is said to be **spontaneous**.
- » It is only dependent on the stability of the particular nuclide. However, if only a single nucleus is considered, it is impossible to predict *when* it will decay. In this way, radioactive decay is **random**.
- » However, we can say that there is a fixed chance of decay occurring within time Δt. Hence, a fixed proportion of a sample containing millions of atoms will decay in that time interval.
- » This randomness is clearly demonstrated by the fluctuations in the count rate observed when radiation from a radioactive isotope is measured with a Geiger counter or other detector.

KEY TERMS

Radioactive decay is a **random** process – it cannot be predicted when a particular nucleus will decay nor which one will decay next. There is a fixed probability that a particular nucleus will decay in any fixed time period.

Radioactive decay is a **spontaneous** process – it is not affected by external factors such as pressure and temperature.

A mathematical treatment of radioactive decay

REVISED ☐

The random nature of radioactive decay means we cannot tell when a particular nucleus will decay, only that there is a fixed chance of it decaying in a given time interval. Thus, if there are many nuclei we can say:

$A = \lambda N$

where A is the **activity**, N is the total number of nuclei in the sample and λ is a constant, known as the **decay constant**.

The activity is measured in a unit called the **becquerel** (Bq).

The decay constant is measured in s^{-1}, min^{-1}, yr^{-1}, etc.

If we measure the activity of an isotope with a relatively short half-life (say 1 minute), we can plot a graph similar to Figure 23.4.

KEY TERMS

The **activity** of a sample of radioactive material is the number of decays per unit time.

The **decay constant** is the probability per unit time that a nucleus will decay.

One **becquerel** is an activity of 1 decay per second.

Check your answers at **www.hoddereducation.com/cambridgeextras**

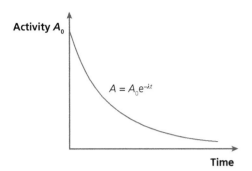

▲ Figure 23.4 Exponential decay of a radioactive isotope

» Study this graph. You will find that the activity falls by equal proportions in successive time intervals.
» The number of atoms decaying in a fixed period decreases because the number of atoms remaining decreases.
» This type of decay is another example of exponential decay (see also pp. 145, 161 and 205) and the equation that describes it is of the form:

$x = x_0 e^{-\lambda t}$

Or in this case:

$A = A_0 e^{-\lambda t}$

» In the example given in the diagram above, the y-axis is labelled activity A, but because the activity is directly proportional to the number of nuclei present (N) it could equally be N.
» The count rate from a detector is also directly proportional to the activity, consequently, this can be used as the variable on the y-axis.

NOW TEST YOURSELF

TESTED ☐

5 a Explain what is meant by the statement, 'Radioactive decay is spontaneous'.
   b Describe the practical evidence for the random nature of radioactive decay.

Half-life and radioactive decay

REVISED ☐

Look at Figure 23.4 again. You have already observed that the activity decreases by equal proportions in equal time intervals. Now look at how long it takes to fall to half the original activity. How long does it take to fall to half this reading (one-quarter of the original)? How long does it take to fall to half of this? You should find that each time interval is the same. This quantity is called the **half-life** ($t_{1/2}$).

Investigating radioactive decay

» When investigating the decay of a radioactive isotope, it is not possible to measure directly either the number of atoms of the isotope remaining in the sample, or the activity of the sample.
» The detector detects only a small proportion of the radiation given off by the sample.
» Radiation is given off in all directions and most of it misses the detector.
» Even radiation that enters the detector may pass straight through it without being detected.
» What is measured is called the **received count rate**.
» The background count rate should be subtracted from the received count rate to give the corrected count rate.

KEY TERMS

The **half-life** of a radioactive isotope is the time taken for the number of undecayed nuclei of that isotope in any sample to reduce to half of its original number.

KEY TERMS

The **received count rate** is defined as the count rate from all sources, as displayed by the detector.

Half-life and the decay constant

If the decay equation is applied to the half-life it becomes:

$$\tfrac{1}{2}N_0 = N_0 e^{-\lambda t_{1/2}}$$

Cancelling the N_0 gives:

$$\tfrac{1}{2} = e^{-\lambda t_{1/2}}$$

Taking logarithms of both sides:

$$\ln(\tfrac{1}{2}) = -\lambda t_{1/2}$$

Therefore:

$$\ln 2 = \lambda t_{1/2}$$

and:

$$t_{1/2} = \frac{\ln 2}{\lambda}$$

Remember that ln 2 is the natural logarithm of 2 and is approximately equal to 0.693. Therefore, the equation may be written as:

$$t_{1/2} = \frac{0.693}{\lambda}$$

WORKED EXAMPLE

The proportion of the carbon-14 isotope found in former living material can be used to date the material. The half-life of carbon-14 is 5730 years. A certain sample has 76.4% of the proportion of this isotope compared with living tissue.

a   Calculate the decay constant for this isotope of carbon.
b   Calculate the age of the material.

Answer

a $\lambda = \dfrac{\ln 2}{t_{1/2}} = \dfrac{0.693}{5730} = 1.21 \times 10^{-4}\,\text{year}^{-1}$

b $N = N_0 e^{-\lambda t}$

$\ln\left(\dfrac{N}{N_0}\right) = -\lambda t$

$t = \dfrac{-\ln(76.4/100)}{1.21 \times 10^{-4}} = 2200\,\text{years}$

NOW TEST YOURSELF

TESTED ☐

6   Table 23.1 shows the received count rate when a sample of the isotope mercury-205 decays. The background count rate is 12 counts per minute.

▼ **Table 23.1**

Time/minutes	0	2.0	4.0	6.0	8.0	10.0
Received count rate/minutes^{-1}	96	78	59	47	40	33
Corrected count rate/minutes^{-1}						

a   Copy and complete the table.
b   Plot a graph of corrected count rate against time.
c   Describe the evidence from these results that radioactive decay is random.
d   Use the graph to determine the half-life of the isotope.
e   Calculate the decay constant of the isotope.

7   The activity of a sample of radioactive tantalum falls by 1% after 40 hours. Calculate:
a   the decay constant
b   the half-life of the isotope

REVISION ACTIVITIES

Two of the important concepts and key terms in this section are mass defect and binding energy. Make sure you understand the link between them and their link to the stability of different nuclides. Go on to explain why energy is released in both fission and fusion. Discuss this with a fellow student – see if you agree with each other.

'Must learn' equations:

$E = mc^2$ $\qquad A = \lambda N$

$N = N_0 e^{-\lambda t}$ $\qquad \lambda = \dfrac{0.693}{t_{1/2}}$

END OF CHAPTER CHECK

In this chapter, you learned to:
» understand the equivalence between energy and mass as represented by the equation $E = mc^2$ ☐
» recall and use the equation $E = mc^2$ ☐
» represent simple nuclear reactions by nuclear equations ☐
» define and use the terms mass defect and binding energy ☐
» sketch the variation in binding energy per nucleon with nucleon number ☐
» explain what is meant by nuclear fusion and nuclear fission ☐
» explain the relevance of binding energy per nucleon to nuclear reactions including nuclear fusion and nuclear fission ☐

» calculate the energy released in nuclear reactions using $\Delta E = \Delta m c^2$ ☐
» understand that fluctuations in count rate provide evidence for the random nature of radioactive decay ☐
» understand that radioactive decay is both spontaneous and random ☐
» define activity and decay constant ☐
» recall and use $A = \lambda N$ ☐
» make a sketch and use the relationship $x = x_0 e^{-\lambda t}$ where x could represent activity, number of undecayed nuclei or corrected received count rate ☐
» define half-life ☐
» use $\lambda = 0.693/t_{1/2}$ ☐
» understand the exponential nature of radioactive decay ☐

Production and use of ultrasound in diagnosis

The piezoelectric effect

REVISED ☐

Ultrasound waves are sound waves that have frequencies above the threshold of human hearing, which is 20 kHz. To produce ultrasound, the **piezoelectric effect** is used.

» Certain materials (for example, quartz) generate an e.m.f. across their crystal faces as a tensile or compressive force is applied.
» Under compression, the e.m.f. is generated in one direction; when under tension the e.m.f. is in the opposite direction.
» This property is used in the piezoelectric microphone.
» A sound wave, which is a pressure wave, can cause a piezoelectric crystal to compress and stretch in a pattern similar to that of the incoming wave.
» This produces a varying e.m.f. across the crystal, which can be amplified as necessary.
» Applying stress produces a voltage, but if a potential difference is applied across such a crystal it either becomes compressed or expands depending on the direction of the e.m.f.

Production of ultrasound

REVISED ☐

» In ultrasound, a single crystal uses the piezoelectric effect to produce and receive the ultrasound waves.
» A short pulse of high-frequency alternating voltage input causes the crystal to vibrate at the same frequency as the input voltage, producing an ultrasonic wave pulse.
» The ultrasonic wave pulse is reflected back from the material under investigation and is received by the same crystal.
» This causes it to vibrate and induce an e.m.f. that is sent to a computer. Thus, the same crystal is both the ultrasound generator and detector.
» This is illustrated in Figure 24.1.
» The term used for a component that acts as both a transmitter and a receiver is a transducer or transceiver.

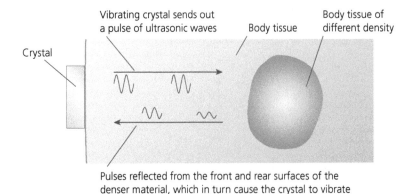

Vibrating crystal sends out a pulse of ultrasonic waves

Body tissue

Body tissue of different density

Crystal

Pulses reflected from the front and rear surfaces of the denser material, which in turn cause the crystal to vibrate

▲ **Figure 24.1 Reflection of ultrasound from tissues of different densities**

> **STUDY TIP**
>
> As with many systems, resonance is required for maximum efficiency: the frequency of the ultrasound should be equal to the natural frequency of vibration of the crystal. For this to occur, the thickness of the crystal must be equal to one-half of the ultrasound wavelength.

Medical use of ultrasound scanning

» Pulses of ultrasound are directed to the organs under investigation.
» Different percentages of incident ultrasound reflect off the boundaries of different types of tissue, enabling a 'sound' picture to be built up.
» As with any waves, the resolution that can be obtained is limited due to diffraction effects.
» Very-high-frequency (therefore short-wavelength) waves are used in medical ultrasound scanning in order to resolve fine details.

> ### NOW TEST YOURSELF
> TESTED ☐
>
> 1 Explain why very-high-frequency ultrasound waves are used in medical scanning.

WORKED EXAMPLE

The approximate speed of sound in human tissue is about $1500 \, \text{m s}^{-1}$. Estimate the minimum frequency ultrasound that would enable a doctor to resolve detail to the nearest $0.1 \, \text{mm}$.

Answer

To resolve detail to the nearest $0.1 \, \text{mm}$, the wavelength of the ultrasound must be of the order of $0.1 \, \text{mm}$.

$$f = \frac{v}{\lambda} = \frac{1500}{0.1 \times 10^{-3}} = 1.5 \times 10^7 = 15 \, \text{MHz}$$

> ### NOW TEST YOURSELF
> TESTED ☐
>
> 2 Calculate the minimum frequency of ultrasound needed to resolve detail of $0.5 \, \text{mm}$ in fatty tissue.
> (speed of sound in fatty tissue = $1.5 \times 10^3 \, \text{m s}^{-1}$)

Acoustic impedance

» When ultrasound moves from a material of one density to another, some of the ultrasound is transmitted and some is reflected.
» The fraction reflected depends on the **acoustic impedance**, Z, of the two materials.

WORKED EXAMPLE

The density of blood is $1060 \, \text{kg m}^{-3}$ and the speed of ultrasound in it is $1570 \, \text{m s}^{-1}$. Calculate the specific acoustic impedance of blood.

Answer

$$Z = \rho c = 1060 \times 1570 = 1.66 \times 10^6 \, \text{kg m}^{-2} \text{s}^{-1}$$

> **KEY TERMS**
>
> Specific **acoustic impedance** (Z) is defined from the equation $Z = \rho c$, where ρ is the density of the material and c is the speed of the ultrasound in the material.

The fraction of the intensity of the ultrasound reflected at the boundary of two materials is calculated from the formula:

$$\frac{I_R}{I_0} = \frac{(Z_1 - Z_2)^2}{(Z_1 + Z_2)^2}$$

where I_R is the intensity of the reflected beam, I_0 is the intensity of the incident beam, and Z_1 and Z_2 are the acoustic impedances of the two materials.

The ratio:

$$\frac{I_R}{I_0}$$

indicates the fraction of the intensity of the incident beam reflected and is known as the **intensity reflection coefficient**.

STUDY TIP

This equation is really only accurate for angles of incidence of 0° but it gives a good approximation for small angles.

WORKED EXAMPLE

The density of bone is $1600\,kg\,m^{-3}$ and the density of soft tissue is $1060\,kg\,m^{-3}$. The speed of sound in the two materials is $4000\,m\,s^{-1}$ and $1540\,m\,s^{-1}$, respectively. Calculate the intensity of the reflected beam compared with the incident beam.

Answer

$Z_{bone} = 1600 \times 4000 = 6.40 \times 10^6\,kg\,m^{-2}\,s^{-1}$

$Z_{soft\ tissue} = 1060 \times 1540 = 1.63 \times 10^6\,kg\,m^{-2}\,s^{-1}$

$\dfrac{I_R}{I_0} = \dfrac{(Z_1 - Z_2)^2}{(Z_1 + Z_2)^2} = \dfrac{(6.40 - 1.63)^2}{(6.40 + 1.63)^2} = \dfrac{4.77^2}{8.03^2} = 0.35$

STUDY TIP

The arithmetic is simplified by ignoring the factor 10^6. It is common to all terms in the equation and therefore cancels.

> ## NOW TEST YOURSELF
TESTED ☐

3 a Calculate the specific acoustic impedance of fatty tissue of density $930\,kg\,m^{-3}$.
   (speed of sound in fatty tissue = $1.5 \times 10^3\,m\,s^{-1}$)

 b Muscle has a specific acoustic impedance of $1.7 \times 10^6\,kg\,m^{-2}\,s^{-1}$. Calculate the intensity reflection coefficient for a fatty tissue–muscle boundary.

4 Calculate the specific acoustic impedance of air.
  (density of air = $1.2\,kg\,m^{-3}$, speed of sound in air = $330\,m\,s^{-1}$)

Coupling medium
REVISED ☐

»» The speed of sound in air is approximately $330\,m\,s^{-1}$ and the density of air is about $1.2\,kg\,m^{-3}$.
»» This gives an acoustic impedance of approximately $400\,kg\,m^{-2}\,s^{-1}$.
»» Comparing this with skin, it means that 99.9% of the incident wave would be reflected at the air–skin boundary.
»» To avoid this, a gel with a similar acoustic impedance to that of skin is smeared on the skin and the ultrasound generator/receiver is moved across this. This gel is known as a **coupling agent** (Figure 24.2).

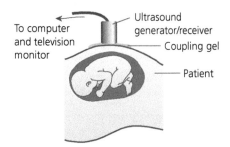

▲ **Figure 24.2 Ultrasound scanning in pregnancy, showing the coupling gel**

Check your answers at **www.hoddereducation.com/cambridgeextras**

A-scan

A pulse of ultrasound is passed into the body and the reflections from the different boundaries between different tissues are received back at the transducer. The signal is amplified and then displayed as a potential difference–time graph on an oscilloscope screen (Figure 24.3).

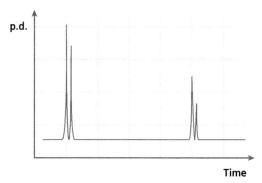

▲ **Figure 24.3 Oscilloscope display of ultrasound pulses**

Figure 24.3 might show the reflections from the front and back of a baby's skull. There are two reflections at each surface – one from the outer part of the skull bone and one from the inner part. Such a scan would give evidence of both the thickness and the diameter of the skull.

WORKED EXAMPLE

Ultrasound travels at a speed of $1500\,\text{m s}^{-1}$ through brain tissue. In Figure 24.3, the time-base of the oscilloscope is set at $50\,\mu\text{s div}^{-1}$. Calculate the diameter of the baby's skull.

Answer

Separation of the two peaks = 4 divisions = $4 \times 50 = 200\,\mu\text{s}$

Distance the ultrasound pulse travels = $vt = 1500 \times 200 \times 10^{-6} = 0.30\,\text{m}$

Diameter of the baby's skull = $\frac{1}{2} \times 0.30 = 0.15\,\text{m}$

» The signal has to travel across the gap between the two sides of the skull, is then reflected and travels back the same distance.
» Hence, the distance calculated from the graph is twice the diameter of the skull.

NOW TEST YOURSELF

5 In an A-scan, the thickness of the bone under test is 52 mm. The reflections from the front and the back of the bone are separated by a time interval of 15 μs. Calculate the speed of sound in bone.

Attenuation of ultrasound

» Ultrasound is absorbed as it passes through tissue.
» The degree of absorption depends on both the type of tissue and on the frequency of the wave.
» The attenuation is exponential and is given by the formula:

$I = I_0 e^{-\mu x}$

where I_0 is the initial intensity of the beam, I is the intensity of the beam after passing through a material of thickness x and μ is the acoustic absorption (or attenuation) coefficient.

WORKED EXAMPLE

A pulse of ultrasound passes through 2.5 cm of muscle and is then reflected back through the muscle to the transducer. Calculate the fraction of the incident beam absorbed by the muscle.

(acoustic absorption coefficient for muscle = 0.23 cm^{-1})

Answer

total thickness of muscle penetrated by ultrasound = $2 \times 2.5 = 5.0$ cm

$$\frac{I_R}{I_0} = e^{-\mu x} = e^{-0.23 \times 5.0} = e^{-1.15} = 0.32$$

fraction absorbed = $1.00 - 0.32 = 0.68$

NOW TEST YOURSELF

TESTED ☐

6   In an ultrasound scan, the ultrasound travels through bone. After passing through 9.6 cm of bone the intensity of the pulse is 52% of the original pulse entering the bone. Calculate the linear acoustic absorption coefficient of bone tissue.

Production and use of X-rays

Production of X-rays

REVISED ☐

» X-rays are formed when electrons are accelerated to very high energies (in excess of 50 keV) and targeted onto a heavy metal (Figure 24.4).
» Most of the electron energy is converted to heat but a small proportion is converted to X-rays.
» The cathode is not actually shown in Figure 24.4. It will be heated indirectly to a high temperature, which causes electrons to be emitted from the surface.
» This is known as **thermionic emission**. The anode is at a much higher potential than the cathode, so the electrons emitted by the cathode are accelerated towards it at very high speeds.

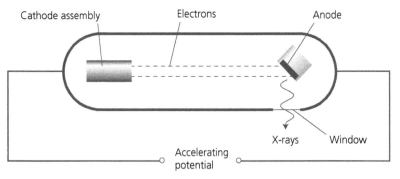

▲ Figure 24.4 Principle of X-ray production

Table 24.1 shows some of the features of a modern X-ray tube.

▼ Table 24.1

Modification	Reason
Rotating anode	To avoid overheating of the anode
Coolant flowing round anode	To avoid overheating of the anode
Thick lead walls	To reduce radiation outside the tube
Metal tubes beyond the window	Collimate and control the width of the beam
Cathode heating control	The cathode in a modern tube is heated indirectly; the current in the heater determines the temperature of the cathode

» The intensity of the X-ray beam is controlled by the number of electrons hitting the anode per unit time.
» The greater the rate of arrival of electrons, the greater is the intensity.
» The rate of arrival of electrons at the anode is controlled by the rate of emission of electrons from the cathode.
» This in turn is controlled by the temperature of the cathode.
» The higher the temperature, the greater is the rate of emission of electrons.

Reducing the dose

A wide range of X-ray frequencies is emitted from a simple X-ray tube. The very soft (long) wavelength rays do not penetrate through the body of the patient, yet would add to the total dose received. An aluminium filter is used to absorb these X-rays before they reach the patient.

X-ray spectrum

A typical X-ray spectrum consists of an emission line spectrum superimposed on a continuous spectrum, as shown in Figure 24.5.

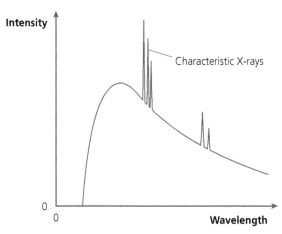

▲ Figure 24.5 A typical X-ray spectrum

» The continuous part of the spectrum is known as Bremsstrahlung or braking radiation.
» It is caused by the electrons interacting with atoms of the target and being brought to rest.
» The large decelerations involved in bringing the electrons to rest produce high-energy photons that are in the X-ray region.

- Sometimes the electrons lose all their energy at once, and at other times give it up through a series of interactions – hence the continuous nature of the spectrum.
- The maximum photon energy cannot be greater than the maximum energy of the incident electrons.
- This means that there is a sharp cut off at the maximum frequency/minimum wavelength of the X-rays.
- The energy of the incident electrons = accelerating voltage × the electronic charge = Ve.
- The maximum energy of the photon emitted = Planck constant × frequency = hf.
- Therefore, for the maximum frequency:

$$hf = Ve$$

and

$$f = \frac{Ve}{h}$$

- Using $c = f\lambda$, the minimum wavelength is:

$$\lambda_{min} = \frac{hc}{Ve}$$

WORKED EXAMPLE

An X-ray tube accelerates electrons through a potential difference of 50 kV before they collide with a tungsten target. Calculate the maximum frequency of the X-rays produced.

Answer

Energy of the electrons, $E = 5.0 \times 10^4\,eV = 5.0 \times 10^4 \times (1.6 \times 10^{-19})\,J = 8.0 \times 10^{-15}\,J$

Energy of the most energetic X-ray = $E = hf = 6.63 \times 10^{-34} \times f_{max}$

Therefore:

$$f_{max} = \frac{8.0 \times 10^{-15}}{6.63 \times 10^{-34}} = 1.2 \times 10^{19}\,Hz$$

NOW TEST YOURSELF

TESTED ☐

7 The minimum wavelength of X-rays from an X-ray tube is 1.5×10^{-12} m. Calculate the potential difference the electrons are accelerated through.

Uses of X-rays in medical imaging

REVISED ☐

Diagnosis of damaged bones and arthritic joints

- A major use of X-rays is in diagnostics, particularly for broken bones and arthritic joints.
- Bone tissue is dense and is a good absorber of X-rays; flesh and muscle are much poorer absorbers.
- Therefore, if a beam of X-rays is incident on an area of damaged bone, a shadow image is formed.
- The bones appear light, because very few X-rays pass through and reach the sensors. The background, where many X-rays reach the sensors, will be much darker.

Diagnosis of ulcerated tissues

» When considering the diagnosis of ulcers, doctors are looking at the absorption by soft tissue, which is much less than for bone.
» Much longer wavelength (or 'softer') X-rays are used.
» Even so, there is little difference in the absorption by healthy tissue and ulcerated tissue.
» In order to improve the **contrast**, the patient is given a drink containing a salt that is opaque to X-rays.
» The ulcerated tissue absorbs more of the salt than healthy tissue and hence absorbs much more of the X-ray radiation.
» This material, often a barium salt, is called a **contrast medium**.

> **KEY TERMS**
>
> The **contrast** is a measure of the difference in brightness between light and dark areas.
>
> A **contrast medium** is a material that is a good absorber of X-rays, which consequently improves the contrast of an image.

Attenuation of X-rays

REVISED

» The formal term for the decrease in intensity of a signal as it passes through a material is attenuation.
» Attenuation of X-rays depends on the material they are passing through – for dense materials such as bone it is high; for less dense materials such as flesh it is much lower.
» However, for each, the attenuation (provided the beam is parallel) is exponential in nature, giving the equation:

$$I = I_0 e^{-\mu x}$$

where I is the intensity, I_0 is the initial intensity, μ is a constant called the **linear attenuation coefficient** (or **absorption coefficient**) and x is the thickness of material the signal has passed through.
» The mathematics of, and dealing with, this equation are identical to that for other exponential decays, such as absorption of ultrasound (p. 205) and radioactive decay (p. 199).
» The equivalent of half-life is the **half-value thickness** (h.v.t.) of the material.
» This is the thickness of material that reduces the intensity of the incident signal to half of its original intensity.
» The linear attenuation coefficient depends not only on the material but also on the hardness of the X-rays that are used.

> **KEY TERMS**
>
> The **linear attenuation coefficient** (or **absorption coefficient**) is a measure of the absorption of X-rays by a material. It depends on both the material and the wavelength of the X-rays.
>
> The **half-value thickness** of a material is the thickness that reduces the intensity of the incident signal to half its original intensity.

WORKED EXAMPLE

Bone has a linear attenuation coefficient of 0.35 cm⁻¹ for X-rays of a particular frequency. Calculate the half-thickness of bone for this type of X-ray.

Answer

$$x_{\frac{1}{2}} = \frac{\ln 2}{\mu} = \frac{0.693}{0.35} = 2.0 \, \text{cm}$$

> ▶ **NOW TEST YOURSELF**   TESTED ☐
>
> 8 Fatty tissue has an attenuation coefficient of 0.9 cm⁻¹. Calculate the percentage of the incident signal that is absorbed when X-rays travel through 5.0 mm of fatty tissue.
> 9 Explain why longer wavelength X-rays are used to diagnose disorders in soft tissues, such as muscle injuries, as opposed to in harder tissues, such as bone.

Computed tomography scanning

The traditional use of X-rays has the major disadvantage of producing only a shadow image. This makes it difficult to identify the true depth of organs. Computed tomography (CT) takes the technology of X-rays a step further. The main principles of CT scanning are:

» The patient lies in the centre of a ring of detectors. An X-ray source moves around the patient, taking many images of the same section of the body at different angles (Figure 24.6).
» The images are put together, using a powerful computer, to form a two-dimensional image of a slice through the patient.
» The patient is moved slightly forward so a two-dimensional image of another slice is made. This is repeated for many slices along a single axis.
» The computer combines the slices together to form a three-dimensional image that can be rotated so that medical practitioners can view the image from different angles.

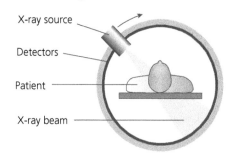

X-ray source

Detectors

Patient

X-ray beam

▲ **Figure 24.6**

> ### NOW TEST YOURSELF
> TESTED
>
> 10 Search the internet and write down three advantages that a CT scan has over a simple X-ray shadow photograph.

Positron emission tomography (PET)

The principles of PET scanning

» PET scanning has a range of uses, such as monitoring cancers, and investigating heart disease, brain function and blood flow.
» It is quite different from ultrasound scanning and X-ray scanning (including CT scanning), where the patient is looked at from the outside.
» In PET scanning, a small amount of radioactive material is injected into a vein.
» It travels around the body and different amounts are absorbed by different tissues.
» This is referred to as a **tracer** or radiotracer.
» Cancerous tissues, in particular, are excellent absorbers of the tracers, consequently tumours show up clearly on a scan.

The tracer commonly used in PET scanning is a β^+ emitter – the isotope fluorine-18, which is tagged onto glucose molecules.

> **KEY TERMS**
>
> A **tracer** (or radiotracer) is a natural compound in which one or more atoms of the natural material are replaced with radioactive atoms of a radioactive isotope of the same element.

> ### NOW TEST YOURSELF
> TESTED
>
> 11 Write down an equation for the decay of fluorine-18.

Annihilation

» The β^+ particle (or positron) is the antiparticle of the electron (p. 95).
» β^+ particles emitted by the tracer travel from the tracer through the body tissues of the patient.
» The positron will only travel a very short distance (less than a millimetre), before interacting with an electron.
» When an electron and the positron interact, the pair annihilate each other.
» At annihilation, all the mass of the pair is converted into two gamma ray photons which travel in opposite directions (180°) to each other (Figure 24.8).

1

Positron   Electron

2

KERPOW!

3

γ-ray photons

▲ Figure 24.8 A cartoon view of energy being released as gamma ray photons on the annihilation of an electron–positron pair

NOW TEST YOURSELF

TESTED

12 Explain why the two γ-ray photons must have equal frequencies and must travel at 180° to each other.
(Hint: think about conservation laws.)

Mathematical treatment of annihilation

WORKED EXAMPLE

Calculate:
a the energy released in an electron–positron annihilation
b the frequency of each of the photons produced
c the wavelength of the photons

Answer

a The mass of a positron is the same as the mass of an electron = 9.1×10^{-31} kg.
When an electron–positron pair annihilate, $2 \times 9.1 \times 10^{-31}$ kg mass becomes the energy of two photons.
Using $E = mc^2 = \frac{1}{2} \times (2 \times 9.1 \times 10^{-31}) \times (3.0 \times 10^8)^2 = 8.19 \times 10^{-14}$ J is the energy of each photon.

b The frequency of the photons is given by $E = hf$, so:
$$f = \frac{E}{h} = \frac{8.19 \times 10^{-14}}{6.63 \times 10^{-34}} = 1.24 \times 10^{20} \, \text{Hz}$$

c $\lambda = \dfrac{c}{f} = \dfrac{3.0 \times 10^8}{1.24 \times 10^{20}} = 2.4 \times 10^{-12} \, \text{m}$

Note: the kinetic energy of the positron is not included in the calculation as it is negligibly small compared to the energy released in annihilation.

NOW TEST YOURSELF　　　　TESTED ☐

13 Calculate the wavelength of the γ-rays produced in a proton–antiproton annihilation.
(mass of a proton = 1.67×10^{-27} kg)

Scanning　　　　REVISED ☐

To scan, the patient is moved through a ring of detectors, as shown in Figure 24.9

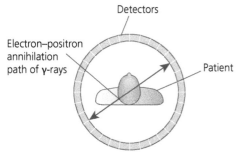

▲ **Figure 24.9 Array of detectors in a PET scanner**

»» The γ-rays produced by the electron–positron annihilation pass through and out of the body and are picked up by the detectors.
»» A powerful computer is used to trace, from which detectors are activated, the line along which the annihilation occurred.
»» The time difference between the detectors being activated enables the computer to determine where along that line the annihilation occurred.
»» The relative numbers of γ-rays coming from each point allows the computer to determine the tracer concentration in each tissue in the patient.

NOW TEST YOURSELF　　　　TESTED ☐

14 Explain why a positron only travels a very short distance before interacting with an electron.

REVISION ACTIVITIES

'Must learn' equations:

$Z = \rho c$

$$\frac{I_R}{I_0} = = \frac{(Z_1 - Z_2)^2}{(Z_1 + Z_2)^2}$$

$I = I_0 e^{-\mu x}$ for absorption for both ultrasound and X-rays

Check through the course to see how often equations of the form $I = I_0 e^{-\mu x}$ occur. Make sure you are really confident working with this type of equation.

END OF CHAPTER CHECK

In this chapter, you have learnt to:

» understand that a piezoelectric crystal changes shape when a p.d. is applied across it and that the crystal generates an e.m.f. when its shape changes ☐

» understand how ultrasound waves are generated and detected by a piezoelectric transducer ☐

» understand how the reflection of pulses of ultrasound at boundaries between tissues can be used to obtain diagnostic information about internal structures ☐

» define the specific acoustic impedance of a medium as $Z = \rho c$, where c is the speed of sound in the medium ☐

» use $I_R/I_0 = (Z_1 - Z_2)^2/(Z_1 + Z_2)^2$ for the intensity reflection coefficient of a boundary between two media ☐

» recall and use $I = I_0 e^{-\mu x}$ for the attenuation of ultrasound in matter ☐

» explain that X-rays are produced by electron bombardment of a metal target and to calculate the minimum wavelength from the accelerating p.d. ☐

» understand the use of X-rays in imaging internal body structures, including an understanding of the term contrast in X-ray imaging ☐

» recall and use $I = I_0 e^{-\mu x}$ for the attenuation of X-rays in matter ☐

» understand the principles of computed tomography (CT) scanning used to produce three-dimensional images of an internal structure ☐

» understand that a tracer (radiotracer) is a substance containing radioactive nuclei that can be introduced into the body and is then absorbed by the tissue being studied ☐

» recall that a tracer that decays by β^+ decay is used in positron emission tomography (PET scanning) ☐

» understand that annihilation occurs when a particle interacts with its antiparticle and that mass–energy are conserved in the process ☐

» explain that, in PET scanning, positrons emitted by the tracer nuclei annihilate when they interact with electrons in the tissue, producing a pair of γ-ray photons ☐

» calculate the energy of γ-ray photons emitted during the annihilation of an electron–positron pair ☐

» understand that the γ-ray photons from an annihilation event travel outside the body and can be detected, and an image of the tracer concentration in the tissue can be created by processing arrival times of the γ-ray photons ☐

25 Astronomy and cosmology

Prior knowledge

To understand astronomy and cosmology, you need to know and understand what is meant by the following terms:

» **star**: a star is a massive ball of ionised gases (a plasma) which are held together by their own gravity; the gases in the star have sufficiently large energies for nuclear fusion reactions to occur, these reactions produce radiant energy which is emitted by the star
» **planet**: a large object that orbits a star
» **solar system**: the star and all the objects that orbit it
» **galaxy**: a group of billions of stars

Standard candles

» Look up at stars in the night sky. You will see that that not all stars appear to be of the same brightness.
» All stars are different, both in size, temperature and in the stage of their evolution.
» The apparent brightness of a star results from the different power emitted by each star and the distance of the star from the observer.
» The total power emitted by a star is known as the **luminosity** of the star. The luminosity of the Sun ($L_\odot$) is 3.83×10^{26} W.
» Consider a source emitting energy equally in all directions.
» When the energy reaches us, only a tiny fraction of the energy emitted by the source enters our eye.
» The further away we are from the source, the more the light spreads out, the smaller the fraction of the emitted energy that enters our eye and the fainter the object appears.
» A useful term to describe the energy incident on unit area is the **radiant flux intensity**, F.
» The unit of radiant flux intensity is the watt per metre squared ($W\,m^{-2}$).

> **KEY TERMS**
>
> The **luminosity** of a star is the **total power** (energy per unit time) emitted by the star.
>
> The **total power** emitted by the source includes not only visible light but the whole electromagnetic spectrum, from radio waves to γ-rays.

> **KEY TERM**
>
> **Radiant flux intensity** (F) is the radiant power passing normally through unit area of a surface.

The inverse square law for radiant flux intensity

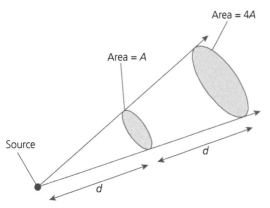

▲ Figure 25.1 A cone of radiation leaving a source

» Figure 25.1 shows how the area illuminated by a cone of light increases as you move away from a source.
» Doubling the distance from the source leads to a doubling of the radius of the illuminated area and hence a quadrupling of the area of the illuminated surface.

> **STUDY TIP**
>
> The mathematics of the inverse square law is identical to the mathematics of both the gravitational field due to a point mass (pp. 121–122) and the electric field due to a point charge (p. 151).
>
> It is worth checking these out.

» The energy emitted from a star is emitted equally in all directions, spreading in an expanding sphere.
» The surface area of a sphere is given by the formula $4\pi r^2$ (where r is the radius of the sphere).
» Hence, at a distance d from a point source, the area of the boundary of the emitted radiation is $4\pi d^2$.
» Consequently, the radiant flux intensity from a star at a distance d from the source is:

$$F = \frac{L}{4\pi d^2}$$

where L is the luminosity of the source.

WORKED EXAMPLE

The distance of the star Proxima Centauri which is relatively close to the Earth is 4.0×10^{16} m. Calculate the radiant flux intensity at Earth from Proxima Centauri.
(luminosity of Proxima Centauri = $1.7 \times 10^{-3} L_\odot$, where $L_\odot$ is the luminosity of the Sun = 3.83×10^{26} W)

Answer

Convert the units of luminosity of Proxima Centauri to watts:

$$L = (1.7 \times 10^{-3}) \times (3.83 \times 10^{26}) = 6.51 \times 10^{23} \text{ W}$$

$$\text{radiant flux intensity at Earth, } F = \frac{L}{4\pi d^2} = \frac{6.51 \times 10^{23}}{4 \times \pi \times (4.0 \times 10^{16})^2}$$

$$F = 3.2 \times 10^{-11} \text{ W m}^{-2}$$

> ## NOW TEST YOURSELF
> TESTED ☐
>
> 1 The distance of the Earth to the Sun is 1.5×10^{11} m. The distance of Mars to the Sun is 2.0×10^{11} m. Compare the radiant flux intensity due to the Sun on Earth and Mars.

Standard candles and determining the distance to galaxies

REVISED ☐

» The term **standard candles** originally goes back to the days when the main source of artificial light was from candles.
» The brightness of different candles could then be compared with candles made in a specific way. These became known as 'standard candles'.
» In astronomy, a star whose luminosity is known is now referred to as a standard candle.
» Astronomers use a type of star called a Cepheid variable as a standard candle. The brightness of a Cepheid variable star varies regularly with a fixed time period; the period depends on the mass of the star and its distance from us.
» Knowing the mass of the star enables the luminosity of the star to be estimated and comparing this with the radiant flux intensity, the distance to the star can be estimated.
» A supernova can also be used as a standard candle for distant galaxies. A supernova occurs when a massive star reaches the end of its life and implodes.
» It is understood that for all supernova of a given type, the maximum peak power output from the implosion is the same, as well as being incredibly large.
» This enables us, using the inverse square law, to calculate the distance to the galaxy in which the supernova occurred.

KEY TERM

A **standard candle** is a class of stellar object that has a known luminosity and whose distance can be determined by calculation using its radiant flux intensity (apparent brightness) and luminosity.

Stellar radii

Black bodies

» You will remember from your earlier studies that an object that is a good absorber of energy is also a good radiator.
» The best possible absorber is one that absorbs 100% of the power incident on it.
» It follows that such an object is also the best possible radiator.
» This theoretical type of object is known as a **black body**.
» No object is a perfect black body; however, many objects approximate to being one and thus black bodies can be used for modelling real objects such as stars.

The Stefan–Boltzmann law

The Stefan–Boltzmann law states that the radiant flux intensity, the power radiated per unit area, of a black body is proportional to the fourth power of its kelvin temperature:

$$F \propto T^4$$

which leads to:

$$F = \sigma T^4$$

where σ is a constant.

The value of $\sigma = 5.7 \times 10^{-8}\,\text{W}\,\text{m}^{-2}\,\text{K}^{-4}$.

The total power emitted by a black body (the luminosity) is the radiant flux intensity emitted by the object multiplied by the surface area of the object. Generally, it is assumed that a star is spherical. The surface area of a sphere = $4\pi r^2$, where r is the radius of the star. This gives:

$$L = 4\pi\sigma r^2 T^4$$

WORKED EXAMPLE

The radius of the Sun is $7.0 \times 10^5\,\text{km}$ and its luminosity is $3.83 \times 10^{26}\,\text{W}$. Calculate the surface temperature of the Sun.

Answer

Rearrange the formula $L = 4\pi\sigma r^2 T^4$ to make T^4 the subject:

$$T^4 = \frac{L}{4\pi\sigma r^2}$$

Convert the radius into metres = $7.0 \times 10^8\,\text{m}$:

$$T^4 = \frac{3.83 \times 10^{26}}{4\pi \times 5.7 \times 10^{-8} \times (7.0 \times 10^8)^2}$$
$$= 1.09 \times 10^{15}\,\text{K}^4$$

Take the 4th root:

$$T = 5700\,\text{K}$$

▶ NOW TEST YOURSELF

2 The surface temperature of the star Betelgeuse is 3600 K, and it has a diameter approximately 800 times greater than the Sun. Calculate the luminosity of Betelgeuse.
(radius of the Sun = $7.0 \times 10^8\,\text{m}$)

Check your answers at **www.hoddereducation.com/cambridgeextras**

Wien's displacement law

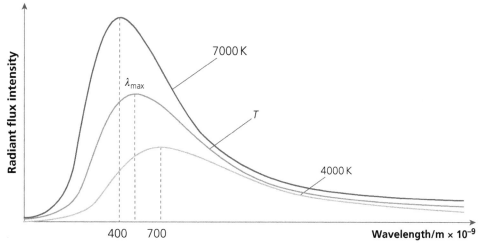

▲ Figure 25.2 The energy spectra of three black bodies at different temperatures

»» Each black body emits a range of wavelengths.
»» The peak of each curve identifies the wavelength that is emitted most strongly by the particular black body. This wavelength is known as the **peak intensity wavelength** and is given the symbol λ_{max}.

Study the graphs in Figure 25.2.

»» The hotter the object, the larger the overall power radiated.
»» The peak intensity wavelength decreases as the temperature of the black body increases.
»» The hotter the black body, the greater the proportion of shorter wavelength (higher frequency and higher photon energy) radiation is emitted.
»» The hottest black body in Figure 25.2 will appear blueish in colour whereas the coolest one will appear red.

> **KEY TERMS**
>
> The **peak intensity wavelength** (λ_{max}) is the wavelength of maximum intensity at a specific temperature.

Wien's displacement law links the peak intensity wavelength to the temperature in a quantitative manner:

$$\lambda_{max} \propto \frac{1}{T}$$

where λ_{max} is the peak intensity wavelength and T is the temperature measured in kelvin.

This can be expressed as an equality by including a constant of proportionality:

$$\lambda_{max} = \frac{C}{T}$$

where C is a constant known as Wien's displacement constant $\approx 2.9 \times 10^{-3}\,\text{m K}$.

The peak intensity wavelength enables scientists to determine the peak surface temperature of a star.

WORKED EXAMPLE

Calculate the peak surface temperature, T, of the star which produced the middle curve in Figure 25.2 and suggest what colour this star will appear to be.

Answer

Reading from the graph, the peak intensity wavelength, λ_{max}, is halfway between the $4.0 \times 10^{-7}\,\text{m}$ and $7.0 \times 10^{-7}\,\text{m}$ marks.

$$\lambda_{max} = 5.5 \times 10^{-7}\,\text{m}$$

Rearrange the formula $\lambda_{max} = \text{constant}\,/T$ to make T the subject:

$$T = \frac{\text{constant}}{\lambda_{max}}$$

$$= \frac{2.9 \times 10^{-3}}{5.5 \times 10^{-7}}$$

$$= 5300\,\text{K}$$

550 nm is in the middle of the electromagnetic spectrum, so the star will appear to be yellow.

3 The star Betelgeuse has a peak surface temperature of approximately 3600 K.
   a Calculate the peak intensity wavelength.
   b The peak intensity wavelength is in the infrared part of the spectrum.
     Explain why the star is visible to the naked eye and suggest what colour it
     appears.

Estimating stellar radii   REVISED ☐

The radius of a star can be estimated if the colour of the star is known and its peak
intensity of radiation is also known. This is then compared with a source of known
radius such as the Sun.

WORKED EXAMPLE

Polaris (sometimes called the Pole Star) has a peak intensity wavelength of
510 nm and a luminosity 2.5×10^3 times the luminosity of the Sun. Estimate the
radius of Polaris.
($L_\odot = 3.83 \times 10^{26}$ W)

Answer

Step 1:

$$\lambda_{max} = \frac{C}{T}$$

Rearrange the equation so that the temperature is the subject:

$$T = \frac{C}{\lambda_{max}}$$

$$= \frac{2.9 \times 10^{-3}}{510 \times 10^{-9}}$$

$$= 5700 \text{ K}$$

Step 2:

$$L = 4\pi\sigma r^2 T^4$$

Rearrange this equation so that r is the subject of the equation:

$$r^2 = \frac{L}{4\pi\sigma T^4}$$

$$= \frac{(2.5 \times 10^3) \times (3.83 \times 10^{26})}{4\pi \times (5.7 \times 10^{-8}) \times 5700^4}$$

$$= 1.27 \times 10^{21} \text{ m}^2$$

$$r = \sqrt{1.27 \times 10^{21}} \text{ m}$$

$$\approx 3.6 \times 10^{10} \text{ m}$$

STUDY TIP

Polaris is a yellow
supergiant star with a
radius about 35 times
greater than the Sun.
The Polaris system is a
triple star with Polaris
itself being by far the
largest of the system
and Polaris A and
Polaris B being yellow
dwarf stars.

NOW TEST YOURSELF   TESTED ☐

4 The star Sirius has a surface temperature of approximately 10 000 K and a
   radius 1.7 times larger than the Sun.
   (temperature of the Sun = 5800 K)
   a Calculate the peak intensity wavelength for Sirius.
   b Calculate the luminosity of Sirius compared with the Sun.

Hubble's law and the Big Bang theory

Absorption line spectra

REVISED ☐

Refer to pp. 190–192 to revise your ideas about emission and absorption line spectra.

» The high temperatures produced in the interior of a star produce a continuous spectrum.
» The light passes through the cooler gases in the star's outer atmosphere.
» Specific frequencies of the radiant energy are absorbed, producing typical absorption line spectra.
» The particular frequencies absorbed are determined by the gases in the outer atmosphere of the star.
» When the absorption line spectra from distant galaxies are observed, it is generally found that the dark lines in the spectra are of longer wavelength (lower frequency) than those formed by the same gas in the Sun.
» This increase in wavelength is known as the **red shift,** because the light is shifted to the red end of the visible spectrum (see Figure 25.3).
» The shift to a longer wavelength indicates that the galaxies are moving away from us, just as when a sound source moves away from an observer a lower pitched note is heard.

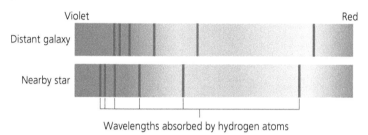

▲ **Figure 25.3 The red shift**

> **STUDY TIP**
>
> The majority of galaxies display a red shift from our viewing platform, the Earth. An exception is the Andromeda Galaxy, our closest neighbouring large spiral galaxy, which demonstrates a shift to the blue end of the spectrum, indicating that Andromeda and the Milky Way are moving towards each other. This 'blue shift' is mainly caused by the gravitational pull between Andromeda and our galaxy, the Milky Way.

» You studied the Doppler effect in sound earlier in the course (p. 61). This is another example of the Doppler effect, this time in electromagnetic radiation.
» Generally, the fainter the galaxy appears, the greater the red shift.
» This suggests that the further away the galaxy is, the greater the speed of recession.
» The decrease in frequency shows that the galaxies are moving away from us. The speed at which they are moving away can be calculated from the change in the frequency, using the formula:

$$\frac{\text{change in wavelength}}{\text{emitted wavelength}} \approx \frac{\text{change in frequency}}{\text{emitted frequency}} \approx$$

$$\frac{\text{speed of recession of the galaxy}}{\text{speed of electromagnetic radiation in a vacuum}}$$

or

$$\frac{\Delta\lambda}{\lambda} \approx \frac{\Delta f}{f} \approx \frac{v}{c}$$

> ## NOW TEST YOURSELF
> TESTED ☐
>
> 5 A dark line in the hydrogen spectrum occurs at a wavelength of 484 nm. When the light from a distant galaxy is analysed, this line has a wavelength of 533 nm. Calculate the speed of recession of this galaxy from the Earth.

Hubble's law

An American scientist, Edwin Hubble, was the first to recognise that Andromeda was not just a group of stars but a galaxy in its own right (1924); our home galaxy was not alone. In his later career, he continued to investigate the distance and the speed of recession of galaxies.

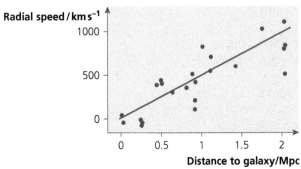

Radial speed / km s^{-1}

Distance to galaxy/Mpc

▲ Figure 25.4 Hubble's data for radial speed against distance

The evidence suggests that the greater the distance to a galaxy, the greater the observed red shift. Hubble went a stage further and suggested that the velocity of recession of galaxies is proportional to their distance from the observer:

$$v \propto d$$

or

$$v \approx H_0 d$$

where H_0 is known as the Hubble constant and has the value $2.4 \times 10^{-18} \, \text{s}^{-1}$.

The approximation sign is used here to emphasise the uncertainties in measuring the distances to different galaxies.

WORKED EXAMPLE

The observed wavelengths of absorption lines in the spectrum from a distant galaxy is increased from the emitted wavelengths by a factor of 0.24. Calculate the speed at which the galaxy is moving away from the observer and the distance it is from the observer.

Answer

Step 1:

$$\frac{\Delta \lambda}{\lambda} = 0.24 \approx \frac{v}{c}$$

Therefore:

$$v \approx 0.24 \times (3.0 \times 10^8)$$

$$= 7.2 \times 10^7 \, \text{m s}^{-1}$$

Step 2:

$$v = H_0 d$$

Rearrange this equation to make d the subject:

$$d = \frac{v}{H_0}$$

$$= \frac{7.2 \times 10^7}{2.4 \times 10^{-18}}$$

$$= 3.0 \times 10^{25} \, \text{m}$$

NOW TEST YOURSELF

6 The helium emission spectrum has a very strong line of wavelength 270 nm. A galaxy is 7.0×10^{25} m from the Earth. Calculate the observed wavelength, on the Earth, of this absorption line.

The Big Bang theory

» It is not only from the Earth that a red shift is seen.
» From any viewing platform, anywhere in the Universe, an observer would see a red shift in the light from other galaxies.
» This suggests that the Universe itself is expanding.

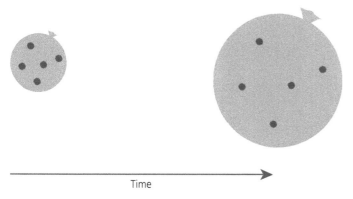

Time

▲ Figure 25.5 A model of the expanding Universe

» The model in Figure 25.5 shows an inflated balloon with dots drawn on it.
» As it is further inflated, all the dots get further away from each other, just as the galaxies in the Universe get further apart.
» This models an expanding Universe.

A thought experiment

Imagine going back in time, millions or billions of years.

» The Universe would be smaller.
» The galaxies would be significantly closer together.

Now go back further, towards the very beginning of time:

» The galaxies would merge together and eventually the Universe itself would become a point.
» At this point, there is nothing and then, almost like an explosion throwing rocks and stones outwards from the centre of the explosion, the Universe is born.
» This is the Big Bang!

> **STUDY TIP**
>
> The term the Big Bang was originally adopted by opponents of the theory in order to belittle it. It is ironic that further evidence has led to their own ideas being abandoned in favour of the Big Bang model.

> ▶ **NOW TEST YOURSELF**
> TESTED ☐
>
> 7 When you stop to think about the Big Bang theory, it is quite difficult to accept. Write, in no more than two sentences, what concerns you about the Big Bang theory.

How old is the Universe?

In the beginning, all the galaxies were at the same point, so our galaxy would have been in the same place as any other galaxy.

Assuming that the speed of recession of a galaxy has been constant since the Big Bang, the time taken to reach its present position relative to the Earth T = distance travelled (d)/speed (v).

But from Hubble's law:

$v = H_0 d$

Therefore:

$T = \dfrac{d}{H_0 d}$

Cancelling:

$$T = \frac{1}{H_0} = \frac{1}{2.4 \times 10^{-18}} = 4.2 \times 10^{17}\,\text{s}$$

This is approximately 13 billion years (1.3×10^{10} years). This is very much an estimate, which is limited by the uncertainties in measuring the distances to different galaxies.

▶ REVISION ACTIVITIES

'Must learn' equations:

$$F = \frac{L}{4\pi d^2}$$

$$\lambda_{max} \propto 1/T$$

$$L = 4\pi\sigma r^2 T^4$$

$$\frac{\Delta\lambda}{\lambda} \approx \frac{\Delta f}{f} \approx \frac{v}{c}$$

$$v \approx H_0 d$$

Look at the internet to find more detail regarding 'standard candles' and the occurrence of blue shift in neighbouring galaxies.

▶ END OF CHAPTER CHECK

In this chapter, you have learnt to:
» understand the term luminosity as the total power of radiation emitted by a star ☐
» recall and use the inverse square law for radiant flux intensity F in terms of luminosity L of the source $F = L/4\pi d^2$ ☐
» understand that an object of known luminosity is called a standard candle ☐
» understand the use of standard candles to determine distance to galaxies ☐
» recall and use the Stefan–Boltzmann law $L = 4\pi\sigma r^2 T^4$ ☐
» recall and use Wien's displacement law $\lambda_{max} \propto 1/T$ to estimate the peak surface temperature of a star ☐
» use Wien's displacement law and the Stefan–Boltzmann law to estimate the radius of a star ☐
» understand that the lines in the emission spectra of distant objects show an increase in wavelength from their known values ☐
» use $\Delta\lambda/\lambda \approx \Delta f/f \approx v/c$ for the red shift of electromagnetic radiation from a source moving relative to an observer ☐
» explain why red shift leads to the idea that the Universe is expanding ☐
» recall and use Hubble's law $v \approx H_0 d$ and explain how this leads to the Big Bang theory ☐

Check your answers at **www.hoddereducation.com/cambridgeextras**

Experimental skills and investigations

Experimental skills are examined in Paper 5, which is worth 30 marks. This is not a laboratory-based paper but, nevertheless, it tests the experimental skills that you will have developed during the second year of your course.

There are two questions on Paper 5, each worth 15 marks. Question 1 tests planning skills. Question 2 tests analysis, conclusions and evaluation. Read the questions carefully and make sure that you know what is being asked of you.

Planning

REVISED

There will be an initial description of some apparatus, which may include a diagram, detailing what quantities can be varied. A formula is given which will enable you to identify which variables are to be investigated. There are four skill areas involved in planning an experiment.

1 Defining the problem

This requires you to look at the task that has been assigned and to write down the variables that impact on the problem:

» the independent variable (the variable that you control)
» the dependent variable (the variable that changes as a result of your changing the independent variable)
» any other variables that might affect the experiment, and which you need to control – generally by attempting to keep them constant

2 Method of data collection

Once you have identified the variables, you have to decide on a method. You have to describe this method clearly and concisely.

» Draw a labelled diagram showing the arrangement of the apparatus.
» Explain how to vary the independent variable.
» Describe how to take measurements of the independent and dependent variables.
» Describe how to control any other variables.
» Consider any safety precautions that are needed.

3 Method of analysis

The third stage is the analysis of results. You must describe how the data collected in stage 2 is to be used.

» Give details of any derived quantities that have to be calculated.
» State which quantities must be plotted on each graph axis.
» Explain how the gradient, intercept and any other quantities derived from the graph are to be used to reach a conclusion.

4 Additional detail and safety considerations

Once the outline of the experiment has been decided, it is important to add the detail necessary to complete the plan. These additional details can be written in stages 1 to 3 and they must include:

» a description of how other variables are kept constant
» details of calculations of derived quantities
» a clear description of measurement techniques including factors such as the range of meters to be used and details of any repeat readings

> » a list of safety considerations
> » a description of sensible precautions necessary to minimise risk

> **STUDY TIP**
>
> **Before the exam:** your practical course should include practice in designing and carrying out experiments. It is only by carrying out experiments that you will learn to look critically at the design and see the flaws in them.
>
> **In the exam:** it is unlikely that you will be given an experiment that you have met before, so do not be put off by something that seems unfamiliar. There will be prompts to guide you in answering the question. It is important that you look at these prompts and follow them carefully.

Analysis, conclusions and evaluation

REVISED ☐

You will be given an equation and a set of data collected from an experiment from which you must find the values of specified constants and estimate the uncertainty in your answer. There are five skill areas.

1 Data analysis

The expression supplied in the question must be rearranged into one of the following forms:

$y = mx + c$

$y = ax^n$

$y = ae^{kx}$

2 Table of results

Using the raw data supplied in a table:

» calculate derived quantities
» include the derived quantities in the table of results with appropriate headings and units
» determine the correct number of significant figures for calculated quantities

3 Graph

Depending on the equation selected in part 1, one of the following graphs is plotted:

» a graph of y against x to find the constants m and c in an equation of the form $y = mx + c$
» a graph of $\lg y$ against $\lg x$ to find the constants a and n in an equation of the form $y = ax^n$
» a graph of $\ln y$ against x to find the constants a and k in an equation of the form $y = ae^{kx}$

When plotting your graph, remember the following:

» error bars must be shown for each point drawn on the graph
» a straight line of best fit is drawn, along with a worst acceptable straight line through the error bars on the graph

4 Conclusion

» Determine the gradient and intercept from the graph.
» Derive relevant expressions that equate to the gradient and intercept.
» Use the expressions to draw relevant conclusions, with the correct units and an appropriate number of significant figures.

Check your answers at **www.hoddereducation.com/cambridgeextras**

5 Treatment of uncertainties

The raw data will have absolute uncertainty estimates for every value. Using these, you should:

>> calculate uncertainty estimates in derived quantities
>> show uncertainty estimates beside every value in the table
>> use the uncertainty estimates to determine the length of error bars on the graph
>> estimate the uncertainty in the gradient and intercept of the graph based on information from differences between the best-fit and worst acceptable straight line
>> estimate uncertainties in the final answer

Examples

Question 1

REVISED ☐

Plan an investigation to determine how the thickness of a particular type of insulation affects the rate of cooling of warm water.

Before starting, you should have in your mind the sort of experiment that you would do to investigate these variables. There is no unique solution. One possibility is to put heated water in a beaker that has insulation wrapped around it and then to measure the rate of cooling of the water.

Stage 1

The independent variable is the thickness of insulation that is used. The dependent variable is the temperature change of the water per unit time. What other variables need to be controlled? Before you read any further, you should jot down some ideas.

Here are some thoughts that you might have considered:

>> maintaining the same mass of water throughout the experiment
>> the temperature of the surroundings should be kept constant
>> the temperature fall during the test should be much smaller than the temperature difference between the container and the surroundings
>> evaporation from the surface of any liquid used should be reduced to a minimum

Stage 2

The next task is to think about how you are going to carry out the experiment. Once you have a method in mind, you need to:

>> describe the method to be used to vary the independent variable
>> describe how the independent variable is to be measured
>> describe how the dependent variable is to be measured
>> describe how other variables are to be controlled
>> describe, with the aid of a clear labelled diagram, the arrangement of apparatus for the experiment and the procedures to be followed
>> describe any safety precautions that you would take

In an experiment to investigate cooling through insulation, you may decide that the simplest way of varying the independent variable is to place the 'test beaker' inside a series of larger beakers and to fill the space between them with the insulating material. The thickness of the insulating material can then be calculated from the diameters of the different beakers and of the test beaker. These diameters could be measured using the internal jaws of a pair of vernier callipers. The cooling rate could be measured by the time taken for a specified drop in temperature.

You should then describe how to ensure that other variables are controlled. You might use a top-pan balance to measure the mass of the test beaker and water between each set of readings. You could ensure that the water is at the same temperature each time by heating it in a constant temperature water bath (and then double checking the temperature before starting the stopwatch).

What extras might you include to ensure that the investigation is as accurate as possible? This tests your experience of doing practical work. Have you sufficient experience to see things that would improve the experiment?

Some of the ideas in the introduction to this part might be included. You might be able to think of some more:

- ›› Choose a temperature drop that is much less than the difference between the starting temperature and room temperature.
- ›› Stir the water in the bath so that it all reaches a uniform temperature.
- ›› Make sure that the water is at the same starting temperature each time.
- ›› Make sure that the room temperature is constant.
- ›› Put a lid on the test beaker to prevent evaporation.
- ›› Use a digital thermometer so that it is easy to spot when the temperature has fallen to a predetermined value.

Finally, a simple labelled diagram of the apparatus is required. This will save a lot of description and can avoid ambiguities.

> **STUDY TIP**
>
> You might find it helpful to write out your description of the stages as a list of bullet points or numbered instructions, rather than as continuous writing. Try it now with this example.

Stage 3

You will probably be told the type of relationship to expect. From this, you should be able to decide what graph it would be sensible to plot. In this example, it might be suggested that the relationship between the variables is of the form:

$$\frac{\Delta T}{\Delta t} = a e^{-kx}$$

where ΔT is the change in temperature in time Δt, x is the thickness of the insulation, and a and k are constants.

Taking natural logarithms of both sides of the equation gives:

$$\ln \frac{\Delta T}{\Delta t} = -kx + \ln a$$

Consequently, if you draw a graph of $\ln (\Delta T/\Delta t)$ against x, it should be a straight line with a negative gradient. The gradient equals k and the intercept on the y-axis is equal to $\ln a$.

Question 2

REVISED ☐

How you tackle this part will depend on which relationship the question asks you to explore. The most likely relationships are of the form $y = a e^{kx}$ or $y = ax^n$.

Relationship $y = ae^{kx}$

To tackle this type of relationship, you need to plot a graph of $\ln y$ against x.

Check your answers at **www.hoddereducation.com/cambridgeextras**

WORKED EXAMPLE

Consider the experiment described in Question 1. This table provides a possible set of results.

Thickness of insulation (x)/cm	Time taken for the temperature to fall 5.0°C/s	Rate of temperature fall $(\Delta T/\Delta t)$/°C s^{-1}	$\ln((\Delta T/\Delta t)/°C\,s^{-1})$
2.0	110	0.0455	−3.090
3.1	127	0.0394	−3.234
4.3	149	0.0336	−3.393
5.1	165	0.0303	−3.497
6.4	195	0.0256	−3.665
7.2	216	0.0231	−3.768

Answer

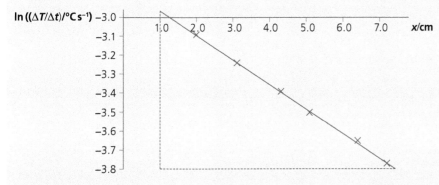

Choose the points (1.0, −2.96) and (7.4, −3.80) to calculate the gradient:

$$\text{gradient} = \frac{\Delta y}{\Delta x} = \frac{-3.80 - (-2.96)}{7.4 - 1.0} = \frac{-0.84}{6.4} = -0.13$$

Hence, $k = -0.13$ cm^{-1}.

To find a, choose a single point and use the generic equation for a straight line, $y = mx + c$.

In this case:

$$\ln(\Delta T/\Delta t) = -kx + \ln a$$

So, choosing the point (1.0, −2.96):

$$-2.96 = -0.13 \times 1.0 + c$$

$$c = -2.83$$

$$c = \ln a$$

Therefore:

$$a = e^c = e^{-2.83} = 0.0590 °C\,s^{-1}$$

Hence, the final expression for the relationship becomes:

$$\frac{\Delta T}{\Delta t} = 0.059e^{-0.13x}$$

Relationship $y = ax^n$

For this type of relationship, you need to draw a graph of $\lg x$ against $\lg y$.

WORKED EXAMPLE 1

An experiment is set up to investigate the diffraction of electrons by a carbon film. The diagram shows the experimental setup.

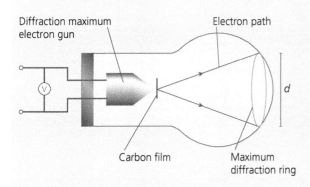

The diameter of the diffraction maximum ring was measured at different accelerating voltages.

The results are recorded in the table.

$V/V \times 10^3$	$d/m \times 10^{-2}$	$\lg(V/V)$	$\lg(d/m)$
2.0	9.8 ± 0.2	3.30	-1.01 ± 0.01
3.0	8.0 ± 0.2	3.48	-1.10 ± 0.01
4.0	6.9 ± 0.2	3.60	-1.16 ± 0.01
5.0	6.2 ± 0.2	3.70	-1.21 ± 0.02
6.0	5.6 ± 0.2	3.78	-1.25 ± 0.01

It is suggested that V and d are related by an equation of the form:

$$d = kV^n$$

where d is the diameter of the maximum ring, V is the accelerating voltage and k and n are constants.

Draw a graph to test the relationship between V and d. Include suitable error bars on your graph.

Draw the line of best fit through the points and also the worst acceptable line. Ensure that the lines are suitably labelled.

Answer

To solve this type of problem a graph of $\lg(V/V)$ against $\lg(d/m)$ is required.

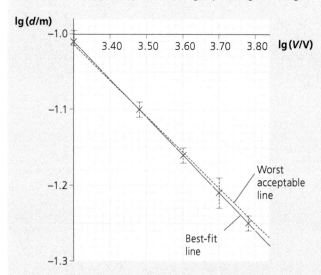

You will observe that the uncertainties for d and for $\lg(d/m)$ are given. In an examination, you would be given the former but you would have to work out the latter. This is done by finding the value of $\lg(d/m)$ for the recorded value and for either the largest or smallest value in the range.

For example, if:

$$d/m = (8.0 \pm 0.2) \times 10^{-2}$$

$$\lg(8.0 \times 10^{-2}) = -1.10 \text{ and } \lg(8.2 \times 10^{-2}) = -1.09$$

$$\text{uncertainty} = 1.10 - 1.09 = 0.01$$

It is important to realise that this process is required for every value of d.

The graph is a straight line, which confirms that the relationship of the form $d = kV^n$ is valid.

STUDY TIP

The worst acceptable line is the line that has either the maximum or the minimum gradient and goes through all the error bars. In Worked example 1, the line of least gradient has been chosen.

Check your answers at **www.hoddereducation.com/cambridgeextras**

WORKED EXAMPLE 2

a Calculate the gradient of the line in Worked example 1. Include the absolute uncertainty in your answer.

b Determine the intercept of the line of best fit. Include the absolute uncertainty in your answer.

c Use your answers to parts a and b to find the values of n and k with their absolute uncertainties

Answer

a For the gradient of the best-fit line, choose the points (3.30, –1.01) and (3.80, –1.26).

$$\text{gradient} = \frac{\Delta y}{\Delta x} = \frac{-1.26 - (-1.01)}{3.80 - 3.30} = \frac{-0.25}{0.50} = -0.50$$

For the worst acceptable line, choose the points (3.30, –1.02) and (3.80, –1.25):

$$\text{gradient} = \frac{\Delta y}{\Delta x} = \frac{-1.25 - (-1.02)}{3.80 - 3.30} = \frac{-0.23}{0.50} = -0.46$$

$$\text{gradient} = -0.50 \pm 0.04$$

b To calculate the intercept, use the generic equation for a straight line:

$$y = mx + c$$

where c is the intercept on the y-axis and m is the gradient = –0.50.
Choose the point (3.30, –1.01):

$$-1.01 = (-0.50 \times 3.30) + c$$

$$c = 0.64$$

To find the uncertainty in c, repeat the procedure using the worst acceptable line. Choose the point (3.30, –1.02), with a gradient –0.46:

$$-1.02 = (-0.46 \times 3.30) + c$$

$$c = 0.498 \approx 0.50$$

The uncertainty in c is 0.64 – 0.50 = 0.14 so c = 0.64 ± 0.14.

c The values of n and k can now be found from the equation:

$$\log d = n \log V + \log k$$

$$y = mx + c$$

So, n is the gradient = –0.50 ± 0.04
$\log k$ is the intercept = 0.64, so $k = 10^{0.64} = 4.4\,(\text{mV}^{-0.5})$

The uncertainty in k is found by using the minimum or maximum value of c to determine the appropriate value of k and finding the difference between that and the intercept.

Minimum value of $c = 0.64 - 0.14 = 0.50$

Minimum value of $k = 10^{0.50} = 3.2$

Difference between intercept and minimum value = $4.4 - 3.2 = 1.2\,\text{mV}^{-0.5}$

The value of $k = (4.4 \pm 1.2)\,\text{mV}^{-0.5}$

From these answers, the full expression for the relationship can be written down:

$$d = 4.4V^{-0.5}$$

which could be written:

$$d = 4.4\sqrt{\frac{1}{V}}$$

Note that the uncertainty in k makes it sensible to round to 2 significant figures.

Exam-style questions

In this section, there are structured questions similar to those you will meet in Paper 4.

You have 2 hours to complete the paper. There are 100 marks on the paper, so you can spend just over 1 minute per mark. If you aim for 1 minute per mark, this will give you some leeway and perhaps time to check through your paper at the end.

See p. 7 for more advice on working through these questions.

Go to **www.hoddereducation.com/cambridgeextras** for example answers and commentaries.

Chapter 12

1 **a** Explain what is meant by a radian. [1]
  b To simulate the large forces that astronauts encounter during the launching of a rocket, they are strapped in a chair attached to an arm of length 12 m. The arm rotates at a frequency of 0.36 Hz.
  i Calculate the time for one revolution of the astronaut. [1]
  ii Calculate the angular speed of the astronaut. [2]
  iii Calculate the force on each kilogram of the astronaut's body. [2]
[Total: 6]

2

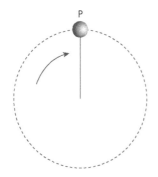

A student rotates a ball on the end of a string in a horizontal circle of diameter 1.9 m. The mass of the ball is 40 g and it travels at a speed of 7.0 m s^{-1}.
  a Explain what is meant by centripetal acceleration. [2]
  b Show that the centripetal acceleration of the ball is 52 m s^{-2}. [2]
  c Calculate the tension in the string. [2]
  d Describe the effect of increasing the mass of the ball on:
  i the centripetal acceleration [1]

  ii the tension in the string [1]
  (Assume the speed remains at 7.0 m s^{-1})
  The student releases the string when the ball is at point P.
  e Draw a line on the diagram to show the path of the ball after the student lets go of the string. [1]
[Total: 9]

Chapter 13

1 **a** Define gravitational potential at a point. [2]
  b A planet has a mass of 6.0×10^{24} kg and radius of 6.4×10^{6} m.
  The planet may be considered to be isolated in space and to have its mass concentrated at its centre.
  i Calculate the gravitational potential at the surface of the planet. [2]
  ii Calculate the energy required to completely remove a spacecraft of mass 800 kg from the planet's surface so that it has sufficient energy to escape from the gravitational field of the planet. You may assume that the frictional forces are negligible. [2]
  iii A single short rocket burn was used for the spacecraft to escape from the surface of the planet to outer space. Calculate the minimum speed that the spacecraft would need to be given by the burn. [3]
[Total: 9]

2 **a** **i** Explain what is meant by a gravitational field. [2]
  ii A planet has a mass M and a radius R. Show that for a satellite orbiting at a height r above the planet its angular speed is given by:
$$\omega^2 = \frac{GM}{(R+r)^3}$$ [2]
  b A weather satellite is placed in a polar orbit (an orbit that goes over the South Pole to over the North Pole and back over the South Pole again). The satellite makes six complete orbits each day. The diameter of the Earth is 12 700 km and its mass is 5.97×10^{24} kg. Calculate:
  i the angular speed of the satellite [2]
  ii the height of the orbit above the surface of the Earth. [3]
[Total: 9]

Chapter 14

1 a Explain what is meant by the term *specific heat capacity*. [2]
   b An iron block of mass 150 g is heated to a temperature of 130°C. It is then lowered into an insulated beaker containing 200 g of water initially at a temperature of 10°C. The temperature of the iron and water reaches a maximum temperature of 19°C.
      i Calculate the specific heat capacity of iron. [4]
         (the specific heat capacity of water = 4200 J kg^{-1}°C^{-1})
      ii State any assumptions you made in the calculation. [1]

[Total: 7]

2 a Explain what is meant by the term *specific latent heat of fusion* of a material. [2]
   b A glass contains 400 g of water at a temperature of 291 K.
      i Determine the temperature of the water in degrees Celsius. [1]
      ii Ice at a temperature of –5°C is added to the water.
         Calculate the minimum amount of ice that must be added so that the water temperature falls to 0°C. [4]
         (specific heat capacity of water = 4200 J kg^{-1}°C^{-1}, specific heat capacity of ice = 2100 J kg^{-1}°C^{-1}, specific latent heat of fusion of ice = 3.3 × 10^5 J kg^{-1})

[Total: 7]

Chapter 15

1 a Explain what is meant by an ideal gas. [2]
   b Argon is stored in a cylinder of volume of 0.20 m^3 at a temperature 0°C and a pressure of 2.0 × 10^6 Pa.
      Determine the number of moles of argon in the cylinder of gas. [2]
   c The argon is used to provide an inert atmosphere in light bulbs. Each bulb has a volume of 120 cm^3. The argon in each bulb exerts a pressure of 5.0 × 10^3 Pa when the temperature is 27°C.
      i Determine the number of moles of argon in each light bulb. [1]
      ii Calculate the number of light bulbs that can be filled with argon from the cylinder. [1]

[Total: 6]

2 a Explain what is meant by the root-mean-square speed of gas molecules. [2]
   b 0.140 m^3 of helium is contained in a cylinder by a frictionless piston. The piston is held in position so that the pressure of the helium is equal to atmospheric pressure and its temperature is 20°C. Assume helium behaves as an ideal gas.
      (atmospheric pressure is 1.02 × 10^5 Pa; mass of helium atom = 6.6 × 10^{-27} kg)
      i Determine the number of moles of helium in the container. [2]
      ii Calculate the average kinetic energy of each helium atom. [2]
      iii Calculate the root-mean-square speed of the helium atoms in the container. [2]
   c The temperature of the helium is gradually increased to 77°C and the helium expands against atmospheric pressure.
      i Calculate the volume of helium at 77°C. [2]
      ii Calculate the total kinetic energy of the helium atoms at 77°C. [2]

[Total: 12]

Chapter 16

1 a Explain what is meant by the internal energy of a system. [2]
   b 1.5 kJ of thermal energy is supplied to a fixed mass of an ideal gas with a volume of 2.0 × 10^{-2} m^3 at a pressure of 1.02 × 10^5 Pa. The gas expands at constant pressure to 2.6 × 10^{-2} m^3. The initial temperature of the gas is 25°C.
      i Calculate the work done by the gas during this expansion. [2]
      ii Calculate the change in internal energy of the gas. [2]
      iii State and explain what happens to the temperature of the gas as it expands. [2]
   c The temperature of the gas gradually returns to its original temperature of 25°C at constant pressure of 1.02 × 10^5 Pa.
      i State and explain whether work is done by the gas or on the gas during this change. [2]
      ii Explain what happens to the internal energy of the gas. [2]

[Total: 12]

Chapter 17

1 The pendulum bob on a large clock has a mass of 0.75 kg and oscillates with simple harmonic motion. It has a period of 2.0 s and an amplitude of 12 cm.

  a i Calculate the angular frequency of the oscillation. [2]

     ii Calculate the maximum restoring force acting on the pendulum bob. [2]

  b i Determine the maximum kinetic energy of the bob. [3]

     ii State the maximum potential energy of the bob. [1]

  c If the clock is not wound up, the oscillation of the pendulum is lightly damped.

     i Explain what is meant by *lightly damped*. [2]

     ii Sketch a graph on the grid to show three periods of this lightly damped oscillation. [2]

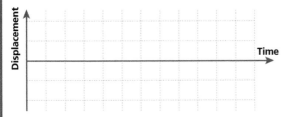

[Total: 12]

2 Tall buildings will oscillate in high winds. The graph shows how the top floor of a particular building oscillates.

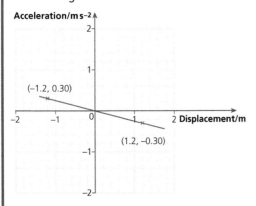

  a i Explain how the graph shows that the building oscillates with simple harmonic motion. [2]

     ii Determine the amplitude of this oscillation. [1]

     iii Calculate the angular frequency of the oscillation. [2]

     iv Determine the period of the oscillation. [2]

     v Calculate the speed at which the building moves when the displacement is 0.60 m from the rest position. [2]

  b Suggest what would happen if an earthquake with a frequency of 0.08 Hz occurred in the region of the building. [2]

[Total: 11]

Chapter 18

1 a Explain what is meant by the electric potential at a point. [1]

  b The spherical dome on a Van de Graaff generator is placed near an earthed metal plate. Consider the dome as an isolated sphere with all its charge concentrated at its centre. The dome has a diameter of 50 cm and the potential at its surface is 65 kV.

     Calculate the charge on the dome. [2]

  c A metal plate is moved slowly towards the dome and the dome partially discharges through the plate, leaving the dome with a potential of 12 kV.

     Calculate the energy that is dissipated during the discharge. [4]

[Total: 7]

2 a A hydrogen atom consists of a proton and an electron. The energy needed to totally remove the electron from the proton (i.e. ionise the atom) is 2.2×10^{-18} J. Calculate the separation of the electron and the proton in the neutral atom. [2]

  b The unit of electrical field strength between two charged parallel plates can be expressed as $N\,C^{-1}$ or as $V\,m^{-1}$. Show that these two units are equivalent. [2]

  c Two horizontal parallel metal plates a distance of 5.0 cm apart in a vacuum have a potential difference of 2.0 kV between them. An electron travelling in a direction parallel to the plates and equidistant between them is moving with a speed of $2.80 \times 10^7\,m\,s^{-1}$.

     i Calculate the electric field strength between the plates. [2]

     ii Calculate the magnitude of the force acting on the electron. [2]

     iii Determine the vertical component of velocity of the electron when it strikes the plate. [3]

     iv Determine the speed of the electron when it strikes the plate. [2]

     v Comment on the difference in trajectory if the electron was replaced by a positron. [1]

[Total: 14]

Check your answers at **www.hoddereducation.com/cambridgeextras**

Chapter 19

1 a Define the capacitance of a parallel plate capacitor. [2]
  b An isolated parallel plate capacitor of capacitance 2200 µF has a potential difference of 12 V across its plates.
    i Calculate the charge on the capacitor. [2]
    ii Calculate the energy stored in the capacitor. [2]
  The capacitor is connected to an identical capacitor as shown.

Capacitor A

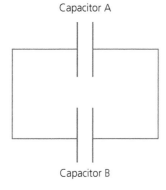

Capacitor B

    iii Calculate the energy now stored in the pair of capacitors. [3]
    iv Suggest a reason for the energy difference between parts ii and iii. [1]
[Total: 10]

2 A capacitor C is connected to a 12 V battery via terminal A of a two-way switch.

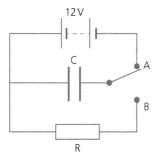

When the switch is in position A, the capacitor is charged. When the switch is moved to terminal B, the capacitor discharges through resistor R. The capacitance of C is 2200 µF and the resistance of R is 2.0 kΩ.
  a Calculate the charge on the capacitor when the switch is in position A. [2]
  The switch is moved to position B at time $t = 0.0$ s.
  b Calculate the time constant RC of this circuit. [2]
  c Sketch a graph of charge on the capacitor against time for the first 12.0 s of the discharge of the capacitor. [3]
  d Calculate the charge on the capacitor 20.0 s after $t = 0.0$ s. [2]
[Total: 9]

Chapter 20

1 a Electrons are accelerated through a potential difference of 4.8 kV. Calculate the velocity of the electrons. [3]
  In a different experiment, electrons travelling at a speed of 2.8×10^7 m s^{-1} enter a uniform magnetic field perpendicularly to the field. The magnetic field has a flux density 4.0 mT.
  b i Explain why the electrons travel in a circular path in the magnetic field. [2]
    ii Calculate the magnitude of the force on the electrons due to the magnetic field. [2]
    iii Calculate the radius of the circular path of the electrons. [2]
  c As the electrons travel through the field, they gradually lose energy. State and explain the effect of this on the radius of the path. [2]
[Total: 11]

2 a State Faraday's law of electromagnetic induction. [2]
  b An aeroplane is flying at a steady altitude in a direction perpendicular to the Earth's magnetic axis. The Earth's magnetic field has a flux density of 34 µT and it makes an angle of 60° with the Earth's surface. The wingspan of the aeroplane is 42 m and it is travelling at a speed of 180 m s^{-1}.
    i Calculate the e.m.f. induced across the wings of the aeroplane. [3]
    ii Explain why this e.m.f. could not drive a current through a conductor connected across the wingtips of the aeroplane. [1]
    iii State and explain the effect on the e.m.f. across the wingtips if the aeroplane were travelling parallel to the Earth's magnetic axis. [2]
[Total: 8]

Chapter 21

1 a A mains supply has a voltage output given by the equation $V = 170 \sin 314t$, where V is the e.m.f. of the supply and t is time in seconds.
    i Calculate the frequency of the supply. [2]
    ii Sketch a graph of the variation of e.m.f. over a time of 50 ms (include values of e.m.f. and time on the axes). [3]
  b An electric toaster that gives a resistive load of 18 Ω is connected across the mains supply.
    i Calculate the r.m.s. voltage of the supply. [2]

ii Determine the mean power output of the toaster. [2]

[Total: 9]

2 Here is a partially completed circuit diagram.

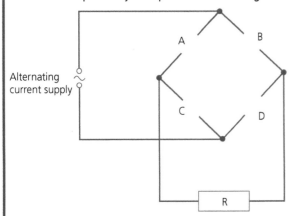

a Copy and complete the diagram by adding diodes at points A, B, C and D to show a complete circuit for full-wave rectification across the load resistor R. [2]

b The alternating current supply has a frequency of 50 Hz and produces a maximum p.d. across R of 100 mV.
Sketch a graph to show two cycles of the p.d. across R (include values of voltage and time). [2]

c The output is used to charge a battery, which requires a smoothed output.
Add a single capacitor to the circuit diagram to give a smoothed output. [2]

d The resistance of the resistor is 1.2 kΩ and the capacitance of the capacitor is 50 μF.
   i Calculate the time constant of this arrangement. [2]
   ii Add lines to the graph in part b to show the smoothed output. [2]
   iii Describe how the output would change if the resistor was replaced by one of resistance 2.4 kΩ. [2]

[Total: 12]

Chapter 22

1 Ultraviolet light of wavelength 300 nm is incident on a zinc surface. The threshold frequency of zinc is 8.8×10^{14} Hz.
   a Explain what is meant by the term *threshold frequency*. [2]
   b i Calculate the photon energy of the ultraviolet light. [2]
   ii Calculate the work function energy of zinc. [2]
   iii Determine the maximum kinetic energy of the photoelectrons emitted when the ultraviolet light is incident on the zinc surface. [2]

iv Calculate the de Broglie wavelength of the photoelectrons emitted with the maximum kinetic energy. [3]

c Explain why no photoelectrons are emitted when the ultraviolet light source is replaced by one that only emits red light. [2]

[Total: 13]

2 The figure represents some of the electron energy levels of a hydrogen atom. When electrons transition between energy levels, photons are absorbed or emitted.

$n = 6$ ———— -0.37 eV
$n = 5$ ———— -0.54 eV
$n = 4$ ———— -0.85 eV
$n = 3$ ———— -1.51 eV

$n = 2$ ———— -3.40 eV

a Explain what is meant by the term photon. [2]

b Calculate the energy required to excite an electron from the $n = 2$ level to the $n = 5$ level. Give your answer in electronvolts and in joules. [2]

c From these energy levels, there are ten possible electron transitions that produce electromagnetic waves.
   i Explain why only four separate colours of light are visible in the emission spectrum from these energy levels. [2]
   ii Identify which transition produces light of wavelength 435 nm. [2]

[Total: 8]

Chapter 23

1 Nuclear power stations use the process of nuclear fission to produce thermal energy.
   a Explain what is meant by the phrase *nuclear fission*. [2]
   b When a uranium-235 nucleus absorbs a neutron, it becomes unstable and breaks down into smaller products. A typical reaction is shown by the following equation:

$$^{235}_{92}\text{U} + ^{1}_{0}\text{n} \rightarrow ^{138}_{55}\text{Cs} + ^{96}_{37}\text{Rb} + 2^{1}_{0}\text{n}$$

The masses of various nuclei and neutrons are given in the table:

Particle	Mass/u
$^{1}_{0}\text{n}$	1.008665
$^{235}_{92}\text{U}$	235.043923
$^{138}_{55}\text{Cs}$	137.911011
$^{96}_{37}\text{Rb}$	95.934284

Check your answers at **www.hoddereducation.com/cambridgeextras**

i Calculate the total mass (in atomic mass units) of the products of this reaction. [1]

ii Determine the mass defect (in atomic mass units) of this reaction. [2]

iii Calculate the energy released (in joules) during a single fission. [2]

c The Sun produces most of its energy by nuclear fusion. Explain what is meant by *nuclear fusion*. [2]

[Total: 9]

2 Protactinium-234 is a radioactive isotope that decays by β-emission with a half-life of 70.2 seconds.

a i Explain what is meant by the term *half-life*. [2]

ii Calculate the decay constant for this isotope. [2]

A student plans to investigate the decay of protactinium-234. She finds that with no radioactive source present, an average count rate of 36 counts per minute is detected. She then places the protactinium-234 source close to the detector and takes three separate readings of the count rate (see table):

Reading	Detected count rate/counts per second
1	81.2
2	72.4
3	51.2

b i Explain why there is a reading on the detector when no radioactive source is present. [2]

ii Comment on the difference in the count rate of the three readings. [2]

iii Assuming the initial reading is taken at time = 0.0 s, what is the time of the third reading? [2]

c i Sketch a graph of detected count rate against time for the first 5 minutes of the decay (include appropriate values on each axis). [3]

ii Predict the count rate 24 hours after the initial reading. [2]

[Total: 15]

Chapter 24

1 a Outline the production and detection of ultrasound as used in ultrasound scanning. [4]

b i Muscle has a specific acoustic impedance of $1.70 \times 10^6 \, \mathrm{kg\,m^2\,s^{-1}}$ and a density of $1080 \, \mathrm{kg\,m^3}$. Calculate the speed of sound through muscle. [2]

ii Ultrasound is incident normally on a muscle–soft tissue boundary.

The specific acoustic impedance of soft tissue is $1.58 \times 10^6 \, \mathrm{kg\,m^2\,s^{-1}}$. Calculate the fraction of the incident ultrasound that is reflected from the boundary. [2]

[Total: 8]

2 a i Explain what is meant by annihilation with reference to a particle–antiparticle interaction. [2]

ii Calculate the wavelength of the γ-rays produced when a neutron–antineutron interact and annihilate. [4]
(mass of a neutron = $1.68 \times 10^{-27} \, \mathrm{kg}$)

b Describe what is meant by a tracer in reference to a PET scan. [2]

c Outline how the position of an electron–positron annihilation is computed in a PET scan. [3]

[Total: 11]

Chapter 25

1 a i Explain what is meant in astronomy by a *standard candle*. [2]

ii Give an example of a standard candle and state the type of information that can be learned from its use. [2]

b A star has a luminosity of $6.4 \times 10^{28} \, \mathrm{W}$. The radiant flux intensity at the Earth is $4.3 \times 10^{-3} \, \mathrm{W\,m^{-2}}$.

i Name the variables that determine the luminosity of a star and state how each of the variables affects the luminosity. [2]

ii Calculate the distance of the star from the Earth. [2]

[Total: 8]

2 a Explain the difference between an emission line spectrum and an absorption line spectrum. [2]

b The Lyman α-line in the hydrogen spectrum has a wavelength of 121.6 nm.

The Andromeda Galaxy is moving towards our galaxy, the Milky Way, at an average speed of about $100 \, \mathrm{km\,s^{-1}}$.

Calculate the change in wavelength of the Lyman α-line in light from Andromeda when viewed from the Earth. [2]

c i Another galaxy is receding from the Earth at $1.8 \times 10^7 \, \mathrm{m\,s^{-1}}$. Use Hubble's law to estimate the distance of this galaxy from the Earth. [2]
(The Hubble constant = $2.4 \times 10^{-18} \, \mathrm{s^{-1}}$)

ii Use Hubble's Law to estimate the age of the Universe in years. State any assumptions you make. [3]

[Total: 9]

Glossary

Key term	Definition	Page number
Absorption coefficient	The **absorption coefficient** (or **linear attenuation coefficient**) is a measure of the absorption of X-rays by a material. It depends on both the material and the wavelength of the X-rays.	209
Acceleration	**Acceleration** is the rate of change of velocity.	17
Accuracy	**Accuracy** is how close to the 'real value' a measurement is.	11
Acoustic impedance	Specific **acoustic impedance** (Z) is defined from the equation $Z = \rho c$, where ρ is the density of the material and c is the speed of the ultrasound in the material.	203
Activity	The **activity** of a sample of radioactive material is the number of decays per unit time.	198
Angular displacement	**Angular displacement** is the change in angle (measured in radians) of an object as it rotates round a circle.	116
Angular speed	**Angular speed** is the change in angular displacement per unit time: $\omega = \Delta\theta/\Delta t$	116
Antineutrino	The **antineutrino** is the antiparticle of the neutrino and is emitted in β^- decay.	97
Antinode	An **antinode** is a point on a stationary wave that has maximum amplitude.	67
Baryons	**Baryons** are a form of hadron which are made up of three quarks (or in the case of antibaryons three antiquarks).	98
Base quantities	**Base quantities** are fundamental quantities whose units are used to derive all other units.	8
Base units	**Base units** are the units of the base quantities.	8
Becquerel	One **becquerel** is an activity of 1 decay per second.	198
Binding energy	The **binding energy** of a nuclide is the work done or energy required to separate to infinity the nucleus into its constituent protons and neutrons.	195
Boltzmann constant	The **Boltzmann constant** $k = R/N_A$, and has the value $1.38 \times 10^{-23}\,\text{J K}^{-1}$.	137
Capacitance	The **capacitance** of a parallel plate capacitor is the charge stored on one plate per unit potential difference between the plates of the capacitor.	156
Centre of gravity	The **centre of gravity** of an object is the point at which the whole weight of the object may be considered to act.	36
Charge	The **charge** passing a point = current × time for which the current flows.	76
Coherent	Wave sources that maintain a constant phase difference are described as **coherent** sources; two or more sources are coherent if they have a constant phase difference.	71
Compression	**Compression** is the difference in length of a spring when a force is applied to the spring and when there is zero load. In a compression, the particles are closer together than normal.	51, 59
Conservation of energy	The law of **conservation of energy** states that the total energy of a closed system is constant.	46
Contrast	The **contrast** is a measure of the difference in brightness between light and dark areas.	209
Contrast medium	A **contrast medium** is a material that is a good absorber of X-rays, which consequently improves the contrast of an image.	209
Coplanar	A set of vectors are said to be **coplanar** if all the vectors in the set lie in the same plane.	13

Check your answers at **www.hoddereducation.com/cambridgeextras**

Decay constant	The **decay constant** is the probability per unit time that a nucleus will decay.	198
Density	The **density** (ρ) of a substance is defined as the mass per unit volume of a substance: ρ = mass/volume	41
Dependent variable	The **dependent variable** is the variable that changes as a result of the changing of the independent variable.	101
Derived units	**Derived units** are combinations of base units.	8
Diffraction	**Diffraction** is the spreading of a wave into regions where it would not spread if it only moved in straight lines after passing through a narrow aperture or past the edge of an object.	71
Displacement	**Displacement** is the distance of an object from a fixed reference point in a specified direction.	17
Distance	**Distance** is the length between two points measured along the straight line joining the two points.	17
Doppler effect	The **Doppler effect** is the change in frequency of waves due to the relative motion of the wave source and the observer.	61
Elastic deformation	**Elastic deformation** means that the object will return to its original shape when the load is removed.	52
Electronvolt	The **electronvolt (eV)** is a unit of energy equal to the energy gained by an electron when it is accelerated through a potential difference of 1 volt.	184
e.m.f.	The **e.m.f.** of a source is numerically equal to the energy transferred (or work done) per unit charge in driving charge around a complete circuit.	83
Energy	**Energy** is defined as the ability (or capacity) to do work.	45
Extension	**Extension** is the difference in length of a spring when there is zero load and when a load is put on the spring.	51
Farad	1 **farad (F)** is the capacitance of a capacitor that has a potential difference of 1 volt across the plates when there is a charge of 1 coulomb on the plates. (1 farad = 1 coulomb per volt)	156
Force	**Force** is the rate of change of momentum.	28
Galvanometer	A **galvanometer** is an instrument for detecting small currents or potential differences. A sensitive ammeter or voltmeter may be used as a galvanometer. They are often used in potentiometers, where a balance point and null reading are being looked for.	90
Gravitational field strength	The **gravitational field strength** at a point is defined as the gravitational force per unit mass at that point. The units of gravitational field strength are $N\,kg^{-1}$.	120
Gravitational potential	The **gravitational potential** at a point is the work done per unit mass in bringing a small test mass from infinity to that point; the units of gravitational potential are $J\,kg^{-1}$.	125
Half-life	The **half-life** of a radioactive isotope is the time taken for the number of undecayed nuclei of that isotope in any sample to reduce to half of its original number.	199
Half-value thickness	The **half-value thickness** of a material is the thickness that reduces the intensity of the incident signal to half its original intensity.	209
Hertz	1 **hertz** is one complete **oscillation** per second.	56
Independent variable	The **independent variable** is the variable that you control or change in an experiment.	101
Intensity	**Intensity** is defined as the energy transmitted per unit time (power) per unit area at right angles to the wave velocity.	58
Internal energy	**Internal energy** is the sum of a random distribution of the kinetic and potential energies associated with the molecules of a system.	138

Internal resistance	**Internal resistance** of a source of e.m.f. is the resistance inherent in the source itself as energy is transferred as charge is driven through the source.	84
Isotopes	**Isotopes** are different forms of the same element which have the same number of protons in the nuclei but different numbers of neutrons.	94
Joule	1 **joule** of work is done when a force of 1 newton moves its point of application 1 metre in the direction of the force.	44
Limit of proportionality	The **limit of proportionality** is the point at which the extension (or compression) of a spring is no longer proportional to the applied load.	51
Linear attenuation coefficient	The **linear attenuation coefficient** (or **absorption coefficient**) is a measure of the absorption of X-rays by a material. It depends on both the material and the wavelength of the X-rays.	209
Load	**Load** is defined as the force that causes deformation of an object.	51
Longitudinal wave	In a **longitudinal wave**, the particles vibrate parallel to the direction of transfer of energy.	59
Luminosity	The **luminosity** of a star is the total power (energy per unit time) emitted by the star. The total power emitted by the source includes not only visible light but the whole electromagnetic spectrum, from radio waves to γ-rays.	214
Magnetic flux	**Magnetic flux** is the product of magnetic flux density and the cross-sectional area perpendicular to the direction of the magnetic flux density.	175
Magnetic flux density	**Magnetic flux density** is defined as the force acting per unit current per unit length on a conductor placed at right angles to the field. Flux density is numerically equal to the force per unit length on a straight conductor carrying unit current at right angles to the field.	167
Magnetic flux linkage	**Magnetic flux linkage** is the product of the magnetic flux passing through a coil and the number of turns on the coil.	175
Mass	**Mass** is the property of an object that resists changes in motion. Mass is a base quantity and its unit, the kilogram, is a base unit.	27
Mass defect	The **mass defect** of a nuclide is the difference between the mass of the nucleus of a nuclide and the total mass of the nucleons of that nuclide, when separated to infinity.	195
Mesons	**Mesons** are hadrons made up of a quark and an antiquark.	98
Moment	The **moment** of a force about a point equals the force multiplied by the perpendicular distance of the line of action of the force from the point.	36
Momentum	**Momentum** (p) is defined as the product of mass and velocity: $p = mv$	27
Monochromatic light	**Monochromatic light** is light of a single frequency and, consequently, a single wavelength.	73
Natural frequency	The **natural frequency** of a vibration is the frequency at which an object will vibrate when allowed to do so freely.	146
Neutrino	The **neutrino** is a particle that is emitted in β+ decay. It has zero charge and zero (or very little) rest mass.	97
Node	A **node** is a point on a stationary wave that has zero displacement.	67
Nucleon number	The **nucleon number** is the total number of protons plus neutrons in a nucleus.	94
Nuclide	A **nuclide** is a single type of nucleus with a specific nucleon number and a specific proton number.	94
Null method	A **null method** is one in which the apparatus is arranged so that a zero reading is required. The zero reading implies that the apparatus is balanced and that the value of an unknown can be found from the values of the constituent parts of the apparatus only.	90

Check your answers at **www.hoddereducation.com/cambridgeextras**

Oscillation	An **oscillation** is one vibration of a particle – for example, from its mean position to the position of maximum displacement in one direction, back to the mean position, then to maximum displacement in the opposite direction and finally back to the mean position.	56
Pascal	1 **pascal** is the pressure exerted by a force of 1 newton acting normally on an area of 1 metre squared.	41
Peak intensity wavelength	The **peak intensity wavelength** (λ_{max}) is the wavelength of maximum intensity at a specific temperature.	217
Plastic deformation	**Plastic deformation** means that the object will *not* return to its original shape when the load is removed.	52
Potential difference (p.d.)	The **potential difference** (p.d.) between two points is numerically equal to the energy transferred (or work done) per unit charge as a test charge moves from one point to the other.	78, 83
Power	**Power** = work done/time taken = energy transformed/time taken; $P = W/t$	48
Precision	**Precision** is the part of accuracy that the experimenter controls by the choice of measuring instrument and the skill with which it is used. It refers to how close a set of measured values are to each other.	11
Pressure	**Pressure** (p) is the normal force per unit area: p = force/area	41
Proton number	The **proton number** is the number of protons in a nucleus.	94
Radian	One **radian** is the angle subtended at the centre of a circle by an arc of equal length to the radius of the circle.	115
Radiant flux intensity	**Radiant flux intensity** (F) is the radiant power passing normally through unit area of a surface.	214
Random	Radioactive decay is a **random** process – it cannot be predicted when a particular nucleus will decay nor which one will decay next. There is a fixed probability that a particular nucleus will decay in any fixed time period.	198
Rarefaction	In a **rarefaction**, the particles are further apart than normal.	59
Received count rate	The **received count rate** is defined as the count rate from all sources, as displayed by the detector.	199
Resonance	**Resonance** is when the natural frequency of vibration of an object is equal to the driving frequency, giving a maximum amplitude of vibration.	68
Root-mean-square (r.m.s.) value	The **r.m.s. value** of the current (or voltage) is the value of direct current (or voltage) that would produce thermal energy at the same rate in a resistor.	180
SI units	**SI units** (Système International d'Unités) are carefully defined units that are used throughout the scientific world for measuring all quantities.	8
Specific heat capacity	The **specific heat capacity** of a material is the energy required to raise the temperature of unit mass of the material by unit temperature.	129
Specific latent heat of fusion	The **specific latent heat of fusion** is the energy required to change unit mass of solid to liquid without change in temperature.	130
Specific latent heat of vaporisation	The **specific latent heat of vaporisation** is the energy required to change unit mass of liquid to vapour without change in temperature.	130
Speed	**Speed** is the distance travelled per unit time.	17
Spontaneous	Radioactive decay is a **spontaneous** process – it is not affected by external factors such as pressure and temperature.	198
Standard candle	A **standard candle** is a class of stellar object that has a known luminosity and whose distance can be determined by calculation using its radiant flux intensity (apparent brightness) and luminosity.	215
Strain	**Strain** is the extension per unit of the unloaded length of the wire.	54
Stress	**Stress** is defined as the normal force per unit cross-sectional area of the wire.	54

Superposition	The principle of **superposition** states that if two waves meet at a point, the resultant displacement at that point is equal to the algebraic sum of the displacements of the individual waves at that point.	66
Terminal potential difference	The potential difference across a source of e.m.f., when there is a current through the source, is known as the **terminal potential difference**.	84
Thermionic emission	**Thermionic emission** is the emission of electrons from a hot metal surface.	206
Threshold frequency	The **threshold frequency** is the minimum-frequency radiation that is required to release electrons from the surface of a metal.	186
Threshold wavelength	The **threshold wavelength** = speed of electromagnetic waves in free space/threshold frequency	186
Torque of a couple	**Torque of a couple** is the magnitude of one of the forces × the perpendicular distance between the lines of action of the forces.	37
Tracer	A **tracer** (or radiotracer) is a natural compound in which one or more atoms of the natural material are replaced with radioactive atoms of a radioactive isotope of the same element.	210
Transverse wave	In a **transverse wave**, the particles vibrate at right angles to the direction of transfer of energy.	59
Uncertainty	**Uncertainty** is the range of values in which a measurement can fall.	11
Uniform object	A **uniform object** means that the centre of gravity of the object is at the geometric centre of the object.	36
Velocity	**Velocity** is the change in displacement per unit time.	17
Wavefront	A **wavefront** is an imaginary line on a wave that joins points that are exactly in phase.	71
Weight	**Weight** is the gravitational pull on an object. Weight is a force and, like all forces, its unit is the newton (N).	27
Work	**Work** is defined as being done when a force moves its point of application in the direction in which the force acts.	44
Work function energy	The **work function energy** is the minimum energy, or minimum work required, to remove an electron from the surface of a metal.	186

Check your answers at **www.hoddereducation.com/cambridgeextras**